Food Systems in an Unequal World

Society, Environment, and Place

Series Editors: Andrew Kirby, Janice Monk, and Paul Robbins

Food Systems in an Unequal World

Pesticides, Vegetables, and Agrarian Capitalism in Costa Rica

Ryan E. Galt

THE UNIVERSITY OF
ARIZONA PRESS
TUCSON

The University of Arizona Press
© 2014 The Arizona Board of Regents
All rights reserved

www.uapress.arizona.edu

Library of Congress Cataloging-in-Publication Data
Galt, Ryan E., 1977–
  Food systems in an unequal world : pesticides, vegetables, and agrarian
capitalism in Costa Rica / Ryan E. Galt.
      pages cm. — (Society, environment, and place)
    Includes bibliographical references and index.
    ISBN 978-0-8165-0603-3 (cloth : alk. paper)
    1. Pesticides—Health aspects—Costa Rica. 2. Pesticides—Government
policy—Costa Rica. 3. Farmers—Health and hygiene—Costa Rica.
4. Farmers—Costa Rica—Social conditions. 5. Agricultural laborers—Health
and hygiene—Costa Rica. 6. Agriculture—Economic aspects—Costa Rica.
I. Title.
  SB950.3.C8G35 2014
  363.738'498097286—dc23

                                2013034410

Manufactured in the United States of America on acid-free, archival-quality
paper containing a minimum of 30% postconsumer waste and processed
chlorine free.

19 18 17 16 15 14   6 5 4 3 2 1

Contents

Illustrations

Acknowledgments

Writing is inevitably a social endeavor—not just because it is communication but also because the author is supported by a large network of people. My largest debt is to the farmers of Northern Cartago and the Ujarrás Valley. Their interest and hospitality was truly remarkable, and I remain grateful and humbled by their generosity. I am also very thankful to the other people in the agrifood system—especially export firm managers and employees, produce buyers, and agrochemical salespeople—who were willing to talk to me. A number of farmers who were especially interested in my study became my guides in their communities as well as good friends: Salomón Montenegro in Buenos Aires de Pacayas, Crisanto Ramírez in Cot, Marcos Sanabria in San Martín de Santa Rosa, Reinaldo Sánchez in Cipreses, and Ignacio Segura in Calle Naranjo de El Yas. *Estoy muy agradecido a ustedes por su ayuda, su tiempo, y su entusiasmo. No tengo palabras para agradecerles.* Two families provided me with a home away from home in Costa Rica, for which I am extremely thankful: the Warn family in San Antonio de Desamparados and the Sánchez family of Cipreses. I also thank the Sanabria family of San Martín de Santa Rosa for their hospitality on so many occasions.

I owe a large debt to my faculty mentors while I was in graduate school—Professors Karl Zimmerer, Matthew Turner, Jamie Peck, Jane Collins, and Bradford Barham—as well as to other influential professors in my education, including Jess Gilbert, Bob Reed, Dick Walker, Michael Watts, Ted Hamilton, Lisa Naughton, Doug Jackson-Smith, and the late Fred Buttel and Bill Freudenburg. As exemplary scholars, their knowledge and advice improved my endeavor immeasurably. Chris Duvall, Dan Mensher, Mara Goldman, Dawn Biehler, Eric Carter, Eric Compas, the University of Wisconsin–Madison Global Studies dissertators' group, Stefanie Hufnagl-Eichiner, Jennifer Blesh, Rob Young, Steven Wolf, Laurie

Drinkwater, Colleen Hiner, Raoul Liévanos, Christie McCullen, Margaret MacSems, Jess Daniel, numerous anonymous reviewers, and journal editors Paul Robbins and Adam Tickell provided important comments and critiques of various pieces of the project at different stages. I am very grateful for the helpful suggestions of the readers of the full manuscript: three anonymous reviewers, Colleen Hiner, Jennifer Blesh, Jess Daniel, and my 2010 political ecology graduate seminar students.

I am fortunate that many organizations deemed this work worth funding. The Cultural and Political Ecology Specialty Group of the Association of American Geographers and the Latin American, Caribbean, and Iberian Studies Program (LACIS) at the University of Wisconsin–Madison funded my initial field research in 2000. Funding for my 2003–2004 fieldwork was provided by the U.S. Department of State's Fulbright IIE Student Fellowship and by the MacArthur Foundation Global Studies Fellowship from the Global Studies Program at the University of Wisconsin–Madison. I also thank the Wisconsin Alumni Research Foundation for the Academic Year Fellowship and LACIS and the U.S. Department of Education for the Foreign Language and Area Studies Fellowship that supported my graduate course work, research, and proposal writing that led to this project. The research presented here would not have been possible without the support of these organizations.

Library and logistical support for this project involved a very large number of helpful people. The geography library staff at the University of Wisconsin–Madison, especially Tom Tews and Richard Swartz, deserve special thanks. In Costa Rica, I wish to thank the many librarians at Biblioteca Nacional, Centro Agronómico Tropical de Investigación y Enseñanza (CATIE), and the other university libraries for help accessing materials, and I also wish to thank the staff of the ADICO community association's retail outlet for their assistance with my peculiar project of documenting pesticide labels. I am very thankful to the many people at the Instituto Nacional de Aprendizaje's organic agriculture school, Unidad Tecnológica en Agricultura Orgánica, who allowed for and assisted me in establishing the experimental plots, even though data from them did not make it directly into the book. At the U.S. Food and Drug Administration, I am thankful to Carolyn Makovi for answering my questions about the application of Environmental Protection Agency regulations in practice. Most important, I thank Allyson Carter at the University of Arizona Press for believing in the book. She as the Press's staff—Scott De Herrera, Amanda Piell, Abby Mogollon, Lela Scott MacNeil, and Leigh McDonald—and Nicholle Lutz at BookComp helped it to become what you, dear reader, have in your hand.

My family has thankfully been a foundation of support for this project. My parents, Trish and Dan Galt, have supported everything I have done professionally with their warmth and encouragement. My wife Eve deserves special and innumerable thanks for supporting the whole process, from project conception to fieldwork to the final writing stages. Oliver, a recent furry and four-legged addition to our family, has thankfully let me know when it is necessary to take a break (for walking, running, or playing). I couldn't ask for two better life companions for this journey.

Food Systems in an Unequal World

Introduction   Pesticide Problems,

Pesticide Paradoxes

Arturo Aguilar[1] stands in front of me in his agrochemical storeroom. I've just arrived at his farm near Paraíso in Cartago Province, Costa Rica. He takes a swig from an unlabeled bottle full of a brown liquid that he periodically sprays on his vegetable and herb crops. I'm horrified, being well versed in the many negative health effects of agrochemical exposure—death from acute poisoning, cancer, birth defects, and undermining of the immune system, to name a few. I fear that Arturo is joining the ranks of hundreds of thousands of small farmers, most notably in India, who have taken their lives because of increasing indebtedness and other negative impacts of the rollback of state support of agriculture (Mohanty 2005). He puts me slightly at ease by grinning loudly and declaring, "I'm an enemy of agrochemicals."[2] Rather than an attempted suicide, Arturo explains that his stunt was to show the lack of toxicity of an organic spray he buys from his farmer friend, who obtains it from EARTH University, a sustainable development-oriented facility in the Caribbean lowlands of Costa Rica.

Like thousands of other farmers, Arturo grows his crops in the highland area of Northern Cartago, a major vegetable-producing area thirty miles east of San José, the capital of Costa Rica. Unlike most of these farmers, he has decided to do his best to reduce and eventually discontinue his use of synthetic pesticides. He is not an organic farmer but belongs to a farmers' organization that sells produce to an agroexport firm servicing North American and European markets and high-end national markets in Costa Rica such as tourist restaurants and supermarkets. The exporter's managers and those at other agroexporters in the area attempt to control farmers' pesticide use by prohibiting the use of the most residual and toxic ones to comply with export market regulations in the United States, Canada, and Europe. Arturo's commitment to reducing his agrochemical use illustrates how, in an increasingly globalized and yet highly unequal world economy, food standards and regulations in the industrialized world affect export-oriented agriculture in developing countries.

By focusing on vegetable production and pesticide use, this book examines how three major processes come together: agrarian capitalism, market regulation in agrifood commodity chains, and the ecological underpinnings of agriculture. At its core, my argument draws on Karl Polanyi's "double movement" of societal self-protection. Polanyi described the ill effects of the commodification or "marketization" of land, labor, food, and money during the development of capitalism in England and Europe and the awakening of society to the social and environmental destruction brought by these tremendous transformations. This created a countermovement that sought to limit the effects of commodification: "human society would have been annihilated but for protective countermoves which blunted the action" of the self-regulating market (Polanyi 1957, 76).

But just as the spread of capitalism has been geographically uneven (Smith 1984), so too have the various countermovements to protect society and the environment from commodification's most negative effects. Here I focus on two regulatory structures—that of the United States and Costa Rica—that are highly unequal in their effectiveness in shaping the agrifood system. By examining the intersection of these two regulatory systems with the dynamics of agrarian capitalism in a specific setting—Northern Cartago and the Ujarrás Valley—I show how market integration produces different socioecological outcomes. These vary according to the strength of the regulatory apparatus and the economic arrangements between produce buyers and farmers. In short, farmers who export to the United States and Europe have lowered their pesticide use and adopted agroecological methods to comply with export requirements, while national market vegetable farmers maintain some of the highest levels of pesticide use in the world. Additionally, export farmers who are better supported—those with a fixed price for their produce—are more likely to use the least toxic and least residual synthetic pesticides and adopt agroecological alternatives to pesticides. The chapters that follow explain why and how these and other important outcomes have occurred.

Understanding the causes of these outcomes offers insights into agrifood system regulation and governance. Yet these kinds of insights are rare these days, in part because of decades of thinking that export-oriented market integration has worse impacts than local or national market integration. But it is also because of a lack of attention. Critical agrifood researchers' focus on industrial agriculture yielded to an almost exclusive focus on alternative food networks where more ecologically oriented agriculture and localized distribution systems dominate. Despite our shift in attention, the dynamics and outcomes of agrochemically dependent agricultural and food systems matter greatly, as they still dominate most lands and diets.

The Paradox of Costa Rican Agriculture

Costa Rica fills the imagination of tourists with protected areas harboring lush forests, monkeys, sandy beaches, sea turtles, and other symbols of tropical nature. But Costa Rica is also a powerhouse of industrial agriculture. Almost all of its production is geared toward export and national markets, as policies supporting subsistence producers were largely phased out in the early 1980s, with structural adjustment imposed by the International Monetary Fund during the debt crisis (Edelman 1999). Costa Rica today produces a great deal of pineapples, bananas, melons, coffee, sugarcane, milk, and beef for export (Secretaría Ejecutiva de Planificación Sectorial Agropecuaria 2011). It is also the third-largest vegetable exporter to the United States after Mexico and Canada (Galt 2010).

Costa Rican agriculture is the world's most pesticide intensive, directly contradicting the country's green reputation. Pesticide intensity is typically measured in weight of pesticides applied over a certain area and over a certain period of time, such as kilograms per hectare per year (kg/ha/year), which I will use throughout the book.[3] Table 0.1 shows Costa Rica as the most pesticide-intensive nation on the planet, at 20.4 kg/ha/year. Latin America averages 6.8 kg/ha/year, while Europe averages 3 kg/ha/year and Africa averages 1.9 kg/ha/year (Food and Agriculture Organization 2004).[4] California, the model of industrial agriculture and often considered the most pesticide-intensive place in the world, used 25.8 kg/ha in 1998 (Kegley, Orme, and Neumeister 2000, 8).

Table 0.1 Pesticide intensity by country, 2001[a]

Latin America	**6.8**
Belize	16.5
Brazil	1.2
Costa Rica	20.4
Dominican Republic	4.5
Ecuador	2.5
Nicaragua	2.4
Paraguay	3.4
Uruguay	3.3
Africa	**1.9**
Burundi	0.1
Cameroon	0.1
Cape Verde	0.1
Eritrea	0.1
Ethiopia	0.1
Guinea-Bissau	0.1
Kenya	0.3
Mauritius	17.7[b]
Rwanda	0.1
Senegal	0.1
Europe	**3.0**
Austria	2.1
Czech Republic	1.4
Denmark	1.4
Estonia	0.5
Finland	0.6
France	4.5
Germany	2.3
Greece	2.8
Hungary	1.1
Iceland	0.9
Ireland	2.0
Italy	6.9
Lithuania	0.2
Malta	11.0
Netherlands	8.0
Norway	0.6

continued

Table 0.1   Continued

Poland	0.6
Portugal	5.5
Romania	0.8
Slovakia	2.1[b]
Slovenia	6.8
Sweden	0.7
Switzerland	3.6
United Kingdom	5.8
Asia & Oceania	1.1
Bahrain	1.3
New Zealand	1.0
Pakistan	0.5
Syrian	0.6[b]
Turkey	1.0
Vietnam	2.3
Average	**3.2**

[a] Data are in kg ai/cultivated ha/yr, except where specified.
[b] Data may be formulated product.
Source: Food and Agriculture Organization (2004); Costa Rica has not reported data to the UN Food and Agriculture Organization since 2001, so this is the latest comparative data available.

Pesticide dependence is so strong in Costa Rica that in 2008 as food riots were taking place in many parts of the world due to rising food commodity prices (Patel and McMichael 2009), the national organization of small- and medium-scale farmers gathered in front of offices of the Ministry of Agriculture and Livestock (known as MAG from its Spanish name, Ministerio de Agricultura y Ganadería) to protest rising pesticide prices. They demanded increased registrations of the generic versions of pesticides to lower their costs compared to the brand-name pesticides created by the transnational agrochemical giants. The farmers argued that cheaper agrochemicals would allow them to better bolster national food security through the production of basic grains that Costa Rican policy had stopped supporting in the 1980s because of neoliberal structural adjustment (Anonymous 2008a). Although much attention focuses on pesticide use on large plantations, this incident illustrates how synthetic pesticides are also extremely important inputs for small- and medium-scale farmers in Costa Rica.

Costa Rica produces a great variety of crops, but of them all, vegetables are the most pesticide intensive. Potato farmers in Northern Cartago, where Arturo lives and works, spray once or twice *per week* with multiple pesticides, meaning that farmers spray pesticide mixtures or "cocktails" about twenty-five times in the four-month potato crop cycle. This is equivalent to 57.3 kg of active ingredient (ai)[5] per hectare per cycle (kg ai/ha/cycle), many times the Costa Rican average. Potato production in Ecuador, located in the Andean center of the crop's origin, averages eight fungicide applications per cycle (Hijmans, Forbes, and Walker 2000, 704). This means that pesticide use on potatoes in Northern Cartago is about three to four times higher than in Ecuador. In the United States, potato farmers use between 16.8 and 28 kg ai/ha/year (Benbrook 1997), while California potato farmers use an average of 40.9 kg ai/ha/year (Kegley, Orme, and Neumeister 2000, 66). These available data suggest that Costa Rican potatoes are one of the world's most heavily sprayed crops.

This is also the case for almost all other vegetables produced in Northern Cartago. Table 0.2 compares Costa Rican pesticide use on fourteen vegetable crops with that of California. On average, Costa Rican vegetables are sprayed with 3.4 times more pesticide than Californian vegetables. Of the fourteen vegetable crops compared, only one—carrots—is more pesticide intensive in California, and this is because of heavy use of dangerous soil fumigants. Because of Costa Rica's high levels of pesticide use, which for vegetables greatly exceeds California's infamous spraying, pesticide reduction efforts in the country deserve attention.

While my purpose here is not to explain Costa Rica's premier place in pesticide use, it is an essential part of the backdrop to this book's story. This backdrop helps me focus instead on a much finer-grained analysis of vegetable production and pesticide use in a particular area and on the broader social and ecological forces shaping them. A fine-grained analysis shows many important lines of difference that point to some of the causal dynamics in the agricultural sector and its regulation.

One difference is between the export and national market—a main topic of this book. A common view of pesticide intensity and markets,

Table 0.2 Intensity of pesticide use on vegetables produced in Costa Rica and California

Crop	Costa Rica pesticide intensity[a]	California pesticide intensity[a]	Ratio of Costa Rica to California
Area average	23.2	25.8	0.9
Beet	58.2	7.7	7.5
Broccoli	18.6	5.7	3.3
Cabbage	31.7	4.0	8.0
Carrot	57.3	83.3	0.7
Cauliflower	33.6	5.0	6.8
Corn	12.0	5.5	2.2
Green bean	12.1	3.7	3.3
Green onion	27.4	12.3	2.2
Lettuce	16.8	10.0	1.7
Onion (sweet)	57.1	14.1	4.1
Peppers	68.5	31.1	2.2
Potato	57.3	40.9	1.4
Squash	16.8	7.5	2.2
Tomato	61.5	33.6	1.8
Vegetable average	37.8	18.9	3.4

[a] kg/ha/year for the area average and kg ai/ha/crop cycle for vegetables.
Source: Author's farmer surveys; Arbeláez and Henao H. (2002, 10–11); Kegley, Orme, and Neumeister (2000, 8, 66–69).

developed by critical social scientists and agroecologists in the 1980s and 1990s, is that export agriculture is more heavily dependent upon agrochemicals than agriculture that provisions local or national markets, largely because they believed that export markets faced more stringent quality requirements. In *Circle of Poison,* Weir and Schapiro (1981, 32) note that "over half, and in some countries up to 70 percent, of the pesticides used in underdeveloped countries are applied to crops destined for export to consumers in Europe, Japan, and the United States. The poor and hungry may labor in the fields, exposed daily to pesticide poisoning, but they do not get to eat the crops protected by pesticides."

Similarly, Barry (1987, 97) states that since the 1950s, "pesticides have become an integral part of cash-crop production in Central America." This could mean cash crops for national consumption, but Barry is referring to only cotton, coffee, and banana exports, since he goes on to state that "of the 11 pounds of pesticides used per capita in the region annually, only a few drops are applied on food crops for local consumption" (104). Thrupp's political ecological work in Central America provides another example. She writes that "the largest amounts and most intensive use of pesticides are in large export-crop plantations of cotton, bananas, coffee, and sugarcane, which form the main basis of these small economies" (Thrupp 1988, 41). More recently, Jorgenson (2007, 75) states that "as farming systems in less-developed countries are integrated into the international economy, often through the influence and control of foreign capital, crop rotation and recycling of organic matter are more likely to be replaced by high-intensity use of pesticides and synthetic fertilizers."

Export crops and foreign capital are the culprits in these narratives. The possibility that national market production—including local capital and the spread of agrarian capitalism endogenously—might lead to similar outcomes is almost entirely ignored, creating an important blind spot. Critical scholars in the 1990s reproduced the same narrative when investigating Latin America's new agroexport boom consisting of high-value horticultural crops such as broccoli and strawberries, which contrasted with traditional agricultural exports of coffee, sugar, bananas, and beef. Andreatta (1998a, 357) notes for the Caribbean generally that the "re-creation of chemical intensive agriculture based on the export of nontraditional crops has led to problems similar to those found in traditional monocultural [export] systems." Murray (1994, 64) states of the new agroexport boom that "as small farmers were transformed from subsistence or local-market producers into export crop farmers, their reliance on the agrochemical technology increased." Thrupp, Bergeron, and Waters (1995, 49) write that with new agroexports, "studies have consistently shown that all kinds of pesticides, including fungicides, insecticides, nematicides and herbicides, are used more intensively for most high-value NTAEs [nontraditional agricultural exports] than for other crops. Doses of pesticide applications per unit land in NTAEs exceed those used on subsistence crops and for crops sold in local markets and are similar or even greater per hectare than in many of the traditional export crops such as coffee and sugarcane."

While I went to the field with these theories, they were challenged by what I encountered. My research in highland Costa Rica instead shows that export crops are generally less pesticide intensive than the same crops grown for the national market. Figure 0.1 shows the average pesticide intensity (in kg ai/ha/cycle) for all vegetable crops for which I gathered data, grouped by market segment.[6] As a whole, national market vegetables are more pesticide intensive than vegetables produced for export: 46.1 kg ai/ha/cycle compared to 19.3 kg ai/ha/cycle. Thus, the conception of pesticide intensity and markets described above—that export crops are pesticide intensive and national market crops are not— is too simplistic. If this binary view held, figure 0.1 would show national market crops falling into the "not pesticide intensive" category, while export crops would be "pesticide intensive" or above. Instead, regardless of market orientation, most vegetable crops grown in the study site fall into the "pesticide intensive" category and above. Potato is the most pesticide-intensive crop, at 11.5 times the national average.[7] Yet crucially, national market crops exhibit wide variation, ranging from "not pesticide intensive" to "extremely pesticide intensive." Nor are export crops uniformly pesticide intensive. While it is true that all export vegetable crops in the area are more pesticide intensive than the Costa Rican average, which reflects a broader trend that vegetables are typically heavily sprayed because they are worth a great deal (Fernandez-Cornejo, Jans, and Smith 1998), data on coffee and sugarcane show that these export crops are not particularly pesticide intensive (for a detailed examination, see Galt 2008b).

Overall, the wide vertical range within the different market segments in figure 0.1 means that the data do not fit the conception of export produce as pesticide intensive and national market produce as not pesticide intensive. The data also do not fit the reverse conception—that export crops are not pesticide intensive and that national market crops are pesticide intensive. The problem with either binary conceptualization is that they are generalized statements about averages that hide considerable variation. They also tell us little about the history and dynamics of these agricultural production systems and how they are tied into wider markets. Below I describe these production systems and markets.

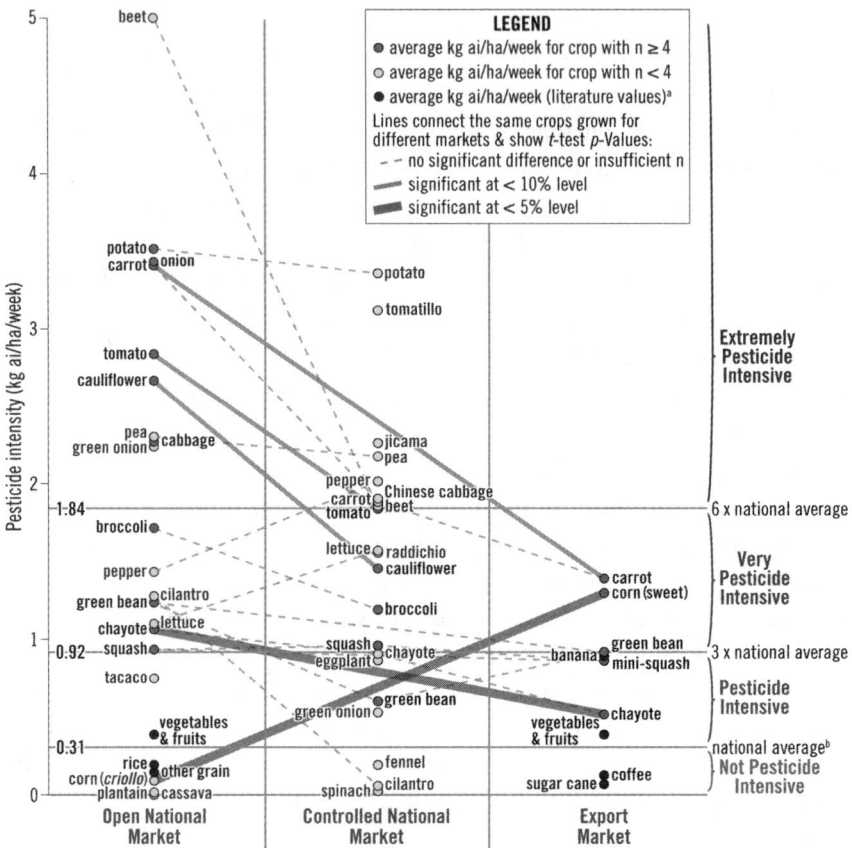

Figure 0.1 — Legend and chart content:

LEGEND
- ● average kg ai/ha/week for crop with n ≥ 4
- ○ average kg ai/ha/week for crop with n < 4
- ● average kg ai/ha/week (literature values)[a]

Lines connect the same crops grown for different markets & show *t*-test *p*-Values:
- – – no significant difference or insufficient n
- ── significant at < 10% level
- ━━ significant at < 5% level

Y-axis: Pesticide intensity (kg ai/ha/week)

Right-side labels:
Extremely Pesticide Intensive
6 x national average
Very Pesticide Intensive
3 x national average
Pesticide Intensive
national average[b]
Not Pesticide Intensive

Column labels: Open National Market / Controlled National Market / Export Market

Sources: Author's farmer surveys 2003-04; [a] from Castillo et al., 1997, p. 42; [b] from Chaverri 1999, p. 6.

Figure 0.1 Vegetable pesticide intensities (kg ai/ha/week) by market compared to the Costa Rican national average and common crops, Northern Cartago and the Ujarrás Valley.

Vegetable Booms in Latin America

By now, consumers in North America are familiar with a seasonless food supply, with the same fresh fruits and vegetables available in grocery stores year-round. A key part of this apparent seasonlessness is Latin America, which supplies many of the fresh fruits and vegetables available in U.S. and Canadian supermarkets in winter. A seasonless food supply is a recent development, depending upon the incredible expansion of long-distance trade in perishable produce. Relative to other perishable foods, fresh vegetables remained more seasonally produced and locally consumed for longer. The globalization of most fresh vegetables did not occur until the 1980s, in part because many vegetables spoiled too quickly and were not worth enough to cover the cost of shipping (Freidberg 2009).

While "fresh" seems an innocent and simple enough term, Freidberg (2009, 2) argues that "far from a natural state," it instead depends "on a host of carefully coordinated technologies, from antifungal sprays to bottle caps to climate-controlled semi trucks." She documents large shifts in fresh as a concept, from depending on a short time and distance from harvest to consumption, to now depending on the technology that protects it: "Now many technologies—from shrink-wrap to irradiation—keep our perishables looking as good as new for longer" (5). This includes the development of the "cold chain," or technologies that control fresh food's climate as it traverses the commodity chain, and high-speed transport that "revolutionized the geography of fresh food" (47).

The privileging of freshness also resulted from changes in how people live and what they value. The most important changes have been around ideas of health and nutrition, with health experts reminding us frequently that "freshness is good for us" (Freidberg 2009, 5). Certain segments of U.S. society—young urban professionals and postcounterculture baby boomers—rejected the mass-produced diet of post–World War II "in favor of presumably healthier, fresher, and more natural foods [that] showed taste and distinction" (191).

This recently expanded demand for fresh produce created a horticultural boom in Latin America starting in the 1980s. Often called the nontraditional agricultural export (NTAE) boom, or what I call new agroexports, these are high-value food exports—a category that includes fresh fruits and vegetables, meats, dairy products, and shellfish—that have in the last few decades unseated the dominance of the global South's traditional (colonial) and more durable export commodities of coffee, cacao, sugar, and tea (Watts and Goodman 1997). As of 1989, high-value foods made up 5 percent of global trade, the same percentage as crude petroleum (Jaffee 1993, 1). The global trade in high-value foods is an important part of the neoliberal, corporate food regime that began in the 1980s (Friedmann and McMichael 1989), of which a key portion is this supermarket revolution for the benefit of "privileged consumers of fresh fruits and vegetables" (McMichael 2009, 142). Scholars lavished critical attention on these new agroexports in the 1990s, focusing on both environmental and social problems.[8]

Yet another much less studied horticultural boom occurred simultaneously. Vegetable production within developing countries has expanded to meet rising domestic urban consumer demand for vegetables, including those grown in hot climates, such as peppers, tomatoes, and eggplant, and those grown in temperate climates, such as potato, carrot, cabbage, onion, cauliflower, broccoli, and lettuce (Dinham 2003; Horst 1987). This vegetable boom has transformed entire landscapes, but studies of vegetables as new agroexports do not typically examine these more nationally based agrarian formations. Since the book focuses on both, I discuss each of these booms in turn below.

The new agroexport boom in Latin America is part of the globalization of fresh produce that exploded in the 1980s. Between 1983 and 1999, U.S. imports of Central American horticultural products, excluding coffee and bananas, rose from $43 million to $456 million, with fruits and vegetables comprising 81 percent of the 1999 imports (Julian, Sullivan, and Sánchez 2000, 1177). This South-North trade is mirrored across the Atlantic, where former African colonies produce fresh fruits and vegetables for their former European colonizers (Dolan and Humphrey 2000; Freidberg 2004; Jaffee 1995).

Costa Rica, Guatemala, and Honduras have been the dominant Central American participants in the new agroexport development strategy, together accounting for 90 percent of new agroexports from the region (Sullivan et al. 1999, 123). Costa Rica took the lead, accounting for 55 percent in 1997, a dominance in part due to a massive expansion in pineapple production. The total value of Costa Rican exports doubled from $1 billion in 1980 to $2 billion in 1993, with the proportion of nontraditional agroexports (everything but coffee, bananas, beef, sugar, and cacao) growing from 38 percent to 60 percent of exports (INICEM-Market Data 1994 and MIDEPLAN 1992, cited in Jansen et al. 1996, 107, table A2.1).

Early on, many researchers painted new agroexports in a glowing light (e.g., von Braun, Hotchkiss, and Immink 1989), while others wrote polemics dismissing them as social and environmental catastrophes (e.g., Conroy, Murray, and Rosset 1996). The major contours of the critique follow those of the Green Revolution and previous agroexport booms. Critics note the likelihood that these new crops would further increase land concentration and social polarization, as in previous agroexport booms (Conroy, Murray, and Rosset 1996; Murray 1994). Thrupp, Bergeron, and Waters (1995, 67) point out that "in most Latin American countries, the main beneficiaries of [new agroexport] growth are large companies, including both transnational corporations (TNCs) and large national and foreign investors." A number of critiques also noted the reduction of nationally oriented production ensured by the geopolitical power of the World Bank and the International Monetary Fund that allowed them to insist on these types of agroexports as part of structural adjustment programs to meet debt obligations (Brohman 1996; Stonich 1993, 1995; Stonich, Murray, and Rosset 1994; Thrupp 1996). A few authors concede mixed and sometimes beneficial results from new agroexport expansion (Carter, Barham, and Mesbah 1995; Hamilton and Fischer 2005; Thrupp, Bergeron, and Waters 1995, 66).

Critics further pointed out that pesticide use is generally high on new agroexports. Pesticide costs for Guatemalan melons for export were $735 to $2,206 per hectare, while export snow peas were greater than $2,206 per hectare (CICP 1988, cited in Thrupp, Bergeron, and Waters 1995, 96). Another study of Guatemalan snow pea production for export showed that 30–35 percent of production costs are for pesticide use (Fisher 1994, cited in Thrupp, Bergeron, and Waters 1995, 96). In export-oriented rose production in Ecuador, 30 percent of production costs go to insecticides and fungicides (Waters 1992, cited in Thrupp, Bergeron, and Waters 1995, 96). This high pesticide use is typically attributed to "strict marketing demands of Northern importers, including: high quality and uniformity of products, with aesthetic criteria for 'perfect' appearance; tight restrictions on export time periods ('windows'); [and] stringent phytosanitary and sanitary standards" (Thrupp, Bergeron, and Waters 1995, 38).

Completely missing from the discussion of new agroexports—and a central contribution of this book—is a comparative perspective that

Table 0.3   Per capita annual fresh produce consumption in Costa Rica, 1991 and 2009

	1991		2009	
	per capita (kg)	rank	per capita (kg)	rank
Fruits				
Banana	8.7	4	17.3	1
Papaya	5.3	5	15.25	2
Pineapple	9.5	3	13.75	3
Apple	—		13.07	4
Orange	12.4	2	11.03	5
Watermelon	—		10.07	6
Mango	—		6.94	7
Strawberry	—		5.72	8
Grapes	—		5.17	9
Plantain	24	1	4.36	10
Melon	—		3.54	11
Jocote	—		3.27	12
Blackberry	—		3	13
Lemon/lime	—		2.86	14
Other fruits	—		20.86	—
Subtotal	*59.9*		*136.19*	
Subtotal, top 5	*59.9*		*70.4*	

continued

simultaneously examines new agroexports and the vegetable production boom for national markets in Latin America. As noted above, this nationally focused boom comes from increased consumer demand. Although the relationship is complex, demand for fresh produce typically goes up (to a point) with increased disposable income (Stewart, Blisard, and Jolliffe 2003). For much of the world, including Costa Rica, this produce remains regionally and nationally produced.

Compared to many Southern nations, Costa Rica has relatively good data available on food consumption. Table 0.3 shows average annual consumption of fresh fruits and vegetables in Costa Rica. Demand for fresh fruits and vegetables surged in recent decades, which can be seen when comparing the consumption of specific foods from 1991 to 2009.

Table 0.3   Continued

	1991		2009	
	per capita (kg)	rank	per capita (kg)	rank
Vegetables, roots, and tubers				
Tomato	9.6	2	12.56	1
Potato	17.1	1	12.16	2
Cabbage	5.1	5	8.84	3
Carrot	5	6	8.14	4
Lettuce	—		8.04	5
Chayote	—		6.53	6
Sweet pepper	1.8	8	5.12	7
Onion	5.9	3	5.02	8
Cilantro	—		4.62	9
Cassava	5.8	4	4.32	10
Cucumber	—		3.42	11
Green bean	—		2.81	12
Broccoli	—		2.51	13
Celery	—		2.21	14
Sweet potato	1.6	9	—	
Tannia	1	10	—	
Cauliflower	2.2	7	—	
Other vegetables	—		14.17	
Subtotal	*55.1*		*100.47*	
Subtotal, top 10	*55.1*		*75.35*	

Source: MAG (1992), cited in Jansen et al. (1996, 57, table 6.1); Programa Integral de Mercadeo Agropecuario (2010, table 11).

Nationally produced crops make up almost all of the fresh vegetables consumed, and the ones in which Northern Cartago and the Ujarrás Valley specialize—potato, tomato, onion, cabbage, carrot, cauliflower, and sweet pepper—dominate national consumption of vegetables. Indeed, the region is the purveyor of a majority of the nation's vegetables, including potatoes.

To feed this demand, as illustrated by Costa Rica, horticultural areas in the global South are increasingly integrated into local and national markets by good roads, which are necessary because of the perishability of fresh produce. This expansion has happened in Costa Rica and many other areas of the developing world.[9] Intensive horticultural production areas typically exist near major urban areas and often in mountain regions, allowing year-round production of temperate crops in tropical latitudes. Horst (1987, 5) describes the expansion of these nationally focused horticultural production systems in Latin America: "Within the past 30 years there have been striking local transformations in the character of traditional agriculture in Latin American tropical highlands between Mexico and Bolivia. A visual survey of the highland rural landscape discloses a widespread expansion in the production of fruits, vegetables, flowers, and bedding and ornamental plants destined for urban markets at home and abroad. This is attributable to rapidly expanding urban populations and a concomitant demand for an array of agricultural products of highland origin."

Compared to Northern researchers' emphasis on export crops (which they may actually eat), these national horticultural production systems as well as other commodity chains that span from developing country to developing country (South-South trade) are not well documented. There is certainly a focus on pesticide problems in national market vegetable production by scholars in Southern nations (Bojacá et al. 2013; Sánchez Saldaña and Betanzos Ocampo 2006; Valdez Salas et al. 2000; see also chapter 6) and focused research on certain crop production systems, such as potato systems in the Andes (Sherwood 2009). But there is little attention to the multiple socioecological facets of these often diverse horticultural production systems, in which farmers grow a number of different vegetables, protect their crops with dozens of pesticides, contract with different produce buyers or selling to intermediaries who sell to spot markets, and are embedded in larger political economic and ecological relationships. The often small scale of the farm operations, year-round production, use of multiple fields with various climates and tenure arrangements, and planting of many crops and varieties for different markets make these farming systems extremely complex and difficult to study and impossible to understand within the tight confines of state agricultural censuses. That these systems also do not fit well with simple assumptions around farm scale and input intensity—since farmers are typically small in scale but rely heavily on agroindustrial inputs—is perhaps another reason why they are not well documented.

There is an urgent need to better understand national market vegetable production systems and their effects on the rural environment, health, and livelihoods. While vegetables are important for a healthy diet, they also are very susceptible to pests and pathogens (Capinera 2001), especially in tropical areas where climate favors pests' rapid reproduction (Hill and Waller 1988). There are many indications that these national horticultural systems are very pesticide intensive. As shown above, in Costa Rica the typical vegetable farmer uses high quantities of pesticides for pest and disease control "to secure his investment in the crop" (Saborío Mora 1994, 400).

Costa Rica is not exceptional in having pesticide-intensive vegetable production for national market. In the area around Yaoundé, Cameroon, farmers of monocultural horticulture systems spend an average of US$190 per hectare on agrochemical inputs, whereas the average for Cameroon is US$6.50 per hectare (Gockowski and Ndoumbe 2004, 199). In these national market horticultural systems in the global south,

> High levels of pesticide use are common, with applications on a weekly basis—or even more often—through the growing season. The insecticides applied are often listed as extremely or highly hazardous (class Ia and Ib) by the World Health Organization, which the Food and Agriculture Organisation of the UN [United Nations] recommends should not be used in developing countries. With the short time-scale from harvest to market, and virtually no possibility of testing for pesticide residues in developing countries, the result is frequently alarmingly high residue levels in urban markets. Ironically some of the healthiest foods could be harbouring dangerous levels of pesticide residues. (Dinham 2003, 575–76)

This book is the first written by an inhabitant of the North that presents a comparative understanding of these two horticultural expansions—the new agroexport boom and the growth in national market vegetable production—and their dynamics as expressed in a specific location. I pay particular attention to how agrarian capitalism, state regulatory systems, and ecological processes jointly shape these systems. Before getting to the specifics of these production systems, we need to see the origins of industrial agriculture and how it has been justified and questioned.

The Questioning of Industrial Agriculture

The most common narratives around pesticide use focus on either their absolute necessity in feeding the world (Avery 1995) or the myriad problems caused by pesticides, as often espoused by environmental groups and critical scholars. The narrative about feeding the world is exceedingly powerful yet extremely naive, as it relies on a vague conceptualization of hunger—it results from an absolute scarcity, not one created through unequal and unjust social relationships—that allows falling back on the grossly simplistic "solution" of producing more food.

This viewpoint has been consistently and forcefully debunked since the 1970s, starting in 1971 with the publication of Frances Moore Lappé's *Diet for a Small Planet* (Lappé 1991) and by subsequent political economic work on hunger and famine (Davis 2001; Poppendieck 1999; Watts 1983b). Critics argue that the strategy of agricultural industrialization and export orientation does not meet the food security needs of the poorest population in developing countries, increases inequality due to socioeconomic scale biases of the crop-technology package, and is not sustainable due to its reliance on heavy agrochemical inputs that cause environmental degradation and serious human health problems (see Lappé et al. 1999). But the persistence of the narrative of feeding the world through agricultural industrialization depends not on its accuracy, since it has been shown to be ignorant of the actual workings of societies and their economies, but rather because it serves powerful interests by reducing fundamentally political problems—how goods and surpluses are distributed in society and who decides on these configurations—to a technological one that seems apolitical. In effect, it is a shell game that shifts the complex question of how we configure society to how we produce more food.

While the narrative of feeding the world through agricultural industrialization is fundamentally problematic, this book does not take it on directly, since that work has been well done elsewhere (Lappé et al. 1999; Perfecto, Vandermeer, and Wright 2009). But I do not align myself with analyses that argue that the benefits of pesticide use outweigh the costs (Avery 1995) or that the dangers have been overblown (Lomborg 2001).

Indeed, as Clarence Cottam (cited in Graham 1970, 63) pointed out decades ago regarding defenses of pesticide use, "Hasn't this side of the problem already been overemphasized by a multi-billion dollar industry employing the most experienced salesmen and lobbyists available?" Instead, following Rachel Carson's (1994) lead, I start with an understanding that highlights pesticide *problems* and the fundamental ecological interconnections between humans and environments.

By now, many consumers are well versed in the problems of industrial agriculture and food systems, such as environmental contamination, reliance on fossil fuels, and increasing food miles. Pesticides remain one of the central concerns that consumers have about the industrialization of their food. Pesticides are chemical compounds that kill or prevent the growth or reproduction of unwanted organisms—insects, fungi, and other pests—that are harmful to humans or organisms valued by humans, typically crops or domesticated animals. Pesticides, synthetic fertilizers applied to the soil, and foliar nutrients applied to plants are all types of agrochemicals.

Various classification systems exist for pesticides. Pesticide classes by mode of action—based on the organisms they kill—are insecticides, acaricides, fungicides, bactericides, herbicides, rodenticides, etc. For this book, fungicides and bactericides are grouped and called "fungicides," while insecticides and acaricides are grouped and called "insecticides." This is a common strategy because many agrochemicals have these dual purposes, and grouping them simplifies discussion considerably. Pesticides are also classified by chemical origin. While somewhat confusing given the new meaning of the word "organic" to mean produced without synthetic chemicals, chemists categorize pesticides into organic or inorganic. Inorganic pesticides are compounds without carbon that are typically created from metallic elements such as arsenic, copper, and lead as well as nonmetallic elements such as sulfur. Here I use the term "inorganic pesticide" in the traditional sense. In contrast, organic chemical compounds—sharing the meaning of organic chemistry—contain carbon and are derived from hydrocarbons found in fossil fuels. In common parlance these compounds of organic chemistry are "synthetic," a use that I adopt here. The term "biopesticide" refers to pesticides of direct biological origin such as plant extracts. Overall, I use the general term "pesticide" to refer to inorganic pesticides, synthetic pesticides, and biopesticides as well as all modes of action—insecticides, fungicides, herbicides, etc.

Pesticides are special anthropogenic chemicals in many ways. Humans create them to intentionally kill organisms, and as a result they are biologically active compounds by design, commonly reacting with many types of life, not just their targets. Unlike most forms of pollution, pesticides are deliberately released into the environment, usually for the perceived short-term economic benefit of the user. In these ways, pesticides contrast with pollution like nuclear and industrial wastes, which are by-products of a production process that yields goods, are not deliberately created to be harmful to life (this is seen as an unintentional side effect), and are not intentionally made to be released into the environment for direct economic benefit (although their very existence assumes that they will be put somewhere, and their release can save money since the costs of disposal become externalized).

Today there exists widespread concern over the ecological and health problems that pesticides cause. Despite widespread concern, pesticide sales remain an important economic activity worldwide. CropLife International (2007, 24–25), the agrochemical-biotechnology industry's representative, reveals that the global market for pesticides—now euphemistically dubbed "conventional crop protection products"—was worth $30.4 billion in 2006, while genetically engineered seeds were worth $6.1 billion (this is the last annual report in which these figures are directly compared). In 2009, global pesticide sales were worth $38 billion (CropLife International 2010, 13). Pesticides and biotechnology are strongly interrelated in that the two major biotech modifications to crops are insect resistance—with the *Bacillus thuringiensis* (Bt) toxin engineered into the plant—and herbicide resistance so that crops can be sprayed with herbicides without killing them.

Inorganic pesticides have been used in agriculture in industrialized nations for more than a century (Whorton 1974), especially in horticulture, which is the growing of fruits, vegetables, and ornamental plants. However, the roots of modern agriculture that depends upon synthetic pesticides can be traced to the birth of the Green Revolution in the 1940s (Wright 1990). The Rockefeller Foundation began the Green Revolution to transform agriculture worldwide in 1941, targeting grains—especially corn, wheat, and rice—for breeding to create high-yielding varieties. These varieties, with larger grain heads and shorter stalks, are high yielding when used together with a new technological package—synthetic fertilizers and pesticides—and under optimal conditions typically involving irrigation. These prerequisites are why they are more accurately called high-response varieties (HRVs).

The Green Revolution was strongly but unsuccessfully contested by Carl Sauer, a geographer at the University of California, Berkeley. Sauer's objection to intervening based on the model of agricultural science at the U.S. land grant universities included the warning that "the example of Iowa is about the most dangerous of all for Mexico. Unless the Americans understand that, they'd better keep out of this country entirely. This thing must be approached from an appreciation of the native economies as being basically sound" (Perkins 1990, 73).

The Rockefeller Foundation and its allies moved ahead nonetheless to transform native economies. In some areas, the results were quite extraordinary, with a dramatic increase in yields of the targeted cereal crops. The Green Revolution became "one of the most important global agents of social and environmental change in the decades after the end of World War II" (Wright 1990, 7). It was important not just because of increased yields but also because it "established an approach and a set of concepts about modernization that have come to dominate agricultural policy, so much so that the term Green Revolution is often used somewhat inaccurately simply to mean agricultural modernization in nonindustrialized countries, or, a little more precisely, the promotion of new seed varieties strongly dependent on synthetic agrochemicals" (Wright 1990, 8).

As a component of the Green Revolution and other forces of agricultural industrialization brought about by market integration, pesticides have increasingly become an integral part of agriculture in many developing countries. Critics of this type of agricultural modernization (Shiva 1991) have focused on this heavy reliance on synthetic pesticides and fertilizers and their numerous negative human and environmental health consequences as well as on the increased social inequalities and upheavals in places where the Green Revolution occurred. Additionally, while it initially appears that pesticides can generate economic benefits for individual farmers, the level at which benefits most accrue is not the farm. Pesticides and new technologies do not necessarily benefit farmers as a whole in capitalist societies because farmers are pitted in competition with one another (Cochrane 1979). This means that pesticide manufacturers and retailers profit most from pesticide use because the risks of production remain with farmers. In other words, it is much better economically to be the input provider rather than the input user.

Pesticides create a number of negative effects for people, society, and the ecosystems on which all life depends (Carson 1994; Pimentel and Greiner 1997). The pesticide problem is complex and systemic because pesticide use negatively impacts human health, the environment, and long-term agricultural productivity and sustainability (Pimentel and Lehman 1993). Supporters of pesticide use often write off critiques of pesticide use as "emotional" or unscientific, a pattern started with Carson's *Silent Spring*. Many of Carson's critics (e.g., Darby 1962) clearly did not understand or did not read the book. In spite of the widespread belief that Carson was entirely antipesticide, she advocated revoking registrations and tolerances (levels of pesticide residues allowed on foods) for two major classes of pesticides: organochlorines and organophosphates. Organochlorines are persistent pesticides that bioaccumulate in organisms and were responsible for the thinning of birds' eggshells in addition to cancer and endocrine disruption leading to other reproductive problems (Colborn, Dumanoski, and Myers 1997). Organophosphates are acutely neurotoxic insecticides related to nerve toxin gases developed during World War II (Russell 2001). Carson (1994, 184) saw it as better "to use less toxic chemicals so that the public hazard from their misuse is greatly reduced. Such chemicals already exist: the pyrethrins, rotenone, ryania, and others derived from plant substances. Synthetic substitutes for the pyrethrins [called synthetic pyrethroids] have recently been developed."

Yet the chemical industry and allied interests vociferously attacked Carson as a mortal enemy (Graham 1970; Hynes 1989), decrying her ecologically based view of human-environment relations. Darby (1962, 60), in his pretentiously paternalistic op-ed titled "Silence, Miss Carson," argued that Carson's perspective meant "the end of all human progress, reversion to a passive social state devoid of technology, scientific medicine, agriculture, sanitation, or education. It means disease epidemics, starvation, misery, and suffering incomparable and intolerable to modern man." Regulating organochlorines and organophosphates has caused none of this to occur, despite unsubstantiated arguments to the contrary that continue to surround DDT (dichlorodiphenyltrichloroethane, an organochlorine insecticide) (Kinkela 2011). Obviously, though, critics aimed not to accurately predict the future but instead to scare the public into accepting the status quo of blanket pesticide applications often of questionable benefit and serious environmental and social cost.

In addition to employing the common positivist fallacy of conflating emotion with ethics, the technocratic perspective argues that societal decisions about pesticides should be based solely on "scientific and economic rationality" through risk assessment and cost-benefit analyses. This position ignores power differentials in setting the terms of value that inform decision making, the self-interest of everyone involved (chemical manufacturers, farmers, farmworkers, governments, and consumers), the different ways that people assess risk, the nature of the risk, and the assessment of alternatives (Nestle 2003; O'Brien 1988, 2000; Patel 2010). Some individuals—especially farmworkers—are heavily exposed to pesticides because of a lack of power (figure 0.2).

The qualitative nature of pesticide risks—"invisible, involuntary, imposed, and uncontrollable" (Nestle 2003, 21)—makes the pesticide debate particularly heated. Without their consent, every person currently living has persistent pesticides such as DDT in their bodies. Robert van den Bosch (1980, 178), an entomologist who pioneered the concept of biological control, pointed out that this "violates our molecular privacy with impunity. . . . Every individual has a right to maximum molecular privacy, and it is society's responsibility to guarantee that right." These exposures generally represent an unknown level of risk because they are new to society (Shrader-Frechette 1985) and raise fears because they can cause dreaded outcomes, including cancer and birth defects.

Pesticide use generates a host of environmental and health problems. Environmental concerns are numerous: groundwater contamination, surface water contamination, bird population declines, fish and wildlife kills, hormonal and developmental disruption of wildlife, mass killings of pollinating bees (including contributing to colony collapse disorder), and reduction in natural enemy populations.[10]

Figure 0.2   A farmworker mixes a batch of pesticides. The ingredients of this mixture include methamidophos (a neurotoxin), mancozeb (a carcinogen), and foliar nutrients for a pesticide application on national market potatoes. The widespread lack of protective equipment use leads to high levels of exposure. (Source: author)

Human health concerns, while sometimes separated from environmental concerns, are always intertwined (Nash 2006), since all humans depend directly or indirectly on the environment for our lives, health, and livelihoods. Concerns about human health include worker and farmer poisoning from spraying and consumers' exposures to pesticide residues on food. Each year, 3 million to 25 million people experience pesticide poisonings, which kill at least 220,000 (Jeyaratnam 1990, 141; World Health Organization 1990, 89). Concerns also extend to environmental exposures in various locations: "The entire risk of the pesticide includes the risks of malfunction of manufacturing and safety equipment in the manufacturing process, and exposure of workers and people in the vicinity of the plant (e.g. Bhopal, India); the risk of accident and spill in storage and transport (e.g. the recent pesticide spill into the Rhine); risks only recently realized to entire aquifers in agricultural states; . . . and risks when used or manufactured in countries where there are minimal safety regulations, or training, or inadequate warning labels in appropriate language" (Hynes 1989, 161).

Epidemiological and toxicological studies point to numerous connections between pesticide exposure and chronic disease. The public tends to think immediately of cancer and birth defects, and indeed numerous cancers have been linked to pesticide exposure (Dich et al. 1997; Flower et al. 2004; Hardell and Eriksson 1999; Osburn 2001; Steingraber 1998; Wesseling et al. 1996). A recent study in Costa Rica, which has the third-highest rate of child leukemia worldwide, has shown that children of parents exposed to pesticides have a significantly higher risk of developing leukemia (Monge et al. 2007). There is evidence for birth defects following exposure to certain pesticides as well (Garry et al. 1996; Moses 1993). Other chronic effects have been uncovered in recent decades, especially the disruption of the endocrine system that governs, among other things, sexual and mental development and the immune system (Barnett et al. 1996; Birnbaum 1994; Colborn 1995; Colborn, vom Saal, and Soto 1993; International Program on Chemical Safety 2002; Krimsky 2000; Porter, Jaeger, and Carlson 1999; Repetto and Baliga 1996; Richard et al. 2005; Vine et al. 2001).

Contemporary food writing has helped make food an increasingly popular interest in the United States, highlighting a growing concern over food-production processes and helping spur the public's interest in social, environmental, and personal consequences of food production and consumption. The rapid growth of the organic food industry and the fair trade, local, and slow food movements demonstrate that consumers increasingly care about the production of their food and what goes into their bodies. Slowly, people are increasingly realizing the sharp contrasts between the virtuous images of food portrayed by the food industry and the often harsh realities of its production.

Of all the negative consequences of pesticide use, the one that generates the most worry among the general population is pesticide residues in food (Baker et al. 2002; Whorton 1974). Demonstrating that low levels of harmful pesticides are either safe or harmful has proven impossible, testing the limits of the sciences of toxicology and epidemiology (Shrader-Frechette 1985; Wing 2000). As Beck (1992, 64) quips, "a central term for 'I don't know either' is 'acceptable level.'" Because the mechanisms by which pesticides affect health are unfamiliar to most citizens, the potential effects are serious but delayed, and the risk is imposed rather than voluntary, pesticides are a dreaded risk (Slovic 1987). Thus, great consumer concern over pesticide residues persists (Knight and Warland 2004).

Given the high amounts of uncertainty, the effects of pesticide residues in food remain vigorously debated. Recently, however, studies have shown that children's main route of exposure to the neurotoxic organophosphate insecticides is through their food and that an organic diet significantly reduces exposure (Lu et al. 2006). Another groundbreaking recent study found that higher prenatal exposure to chlorpyrifos, a commonly used organophosphate insecticide in the United States and Costa Rica, as measured in umbilical cord blood plasma is significantly associated with seven-year-old children's poorer performance on cognitive tests, including IQ and working memory (Rauh et al. 2011). Although this study presents some of the first human-based results of neurological impairment from low-dose exposures, it is not surprising given that studies of developing animals show that low-dose organophosphate exposures impair neurodevelopment and growth (Eskenazi, Bradman, and Castorina 1999).

The fear of pesticide residues is understandable because of the type of risk. Since "we truly are what we eat, food raises questions of intimacy and identity and provokes feelings of anxiety" (Nestle 2003, 20). Additionally, political struggles over the pesticide problem are intense because of the uneven distribution of costs and benefits that are shaped by considerable injustice. The fact that the highest costs are borne most by the rural poor (Bull 1982; Thrupp 1991b)—especially farmworkers and their partners and children—makes pesticide use an important environmental justice issue (Bullard 1993; Chávez 1993; Moses 1993; Pulido 1996; Pulido and Pena 1998). A prominent argument that couples critique of pesticide injustices with an understanding of vast inequalities in the world is *Circle of Poison,* which

> documents a scandal of global proportions—the export of banned pesticides from the industrial countries to the third world. . . . Dozens of pesticides too dangerous for unrestricted use in the United States are

shipped to underdeveloped countries. There, lack of regulation, illiteracy, and repressive working conditions can turn even a "safe" pesticide into a deadly weapon. According to the World Health Organization, someone in the underdeveloped countries is poisoned by pesticides every minute.

But we are victims too. Pesticide exports create a circle of poison, disabling workers in American chemical plants and later returning to us in the food we import. Drinking a morning coffee or enjoying a luncheon salad, the American consumer is eating pesticides banned or restricted in the United States, but legally shipped to the third world. (Weir and Schapiro 1981, 3)

While this framing is increasingly less applicable since social movements and the dynamics of capitalism have changed many of the circumstances (Galt 2008a; Wright 1986), it remains a powerful narrative and is a common take on how inequalities intersect with pesticide hazards.

Weighing the costs and benefits of pesticide use is a difficult proposition. Indeed, even agreeing on the best approach is problematic, as cost-benefit analysis embeds utilitarian ethics that downplay the consequences of unequal power. One study examining the costs and benefits of pesticide use in the United States estimates that the total costs of pesticide use to U.S. society were $9 billion per year, while farmers spent $10 billion on pesticides and saved $40 billion on their crops annually (Pimentel 2009, 106). The $9 billion in annual costs includes both the $3 billion in costs borne jointly by farmers and society—natural enemy destruction and pests' resistance to pesticides—and the $6 billion paid by society in environmental and health damage, including acute and chronic health effects (including cancer); water pollution; bee and domestic animal poisonings; fishery, wild bird and animal losses; and government expenses for controlling pesticide contamination.

Thus, while pesticides, and conventional agriculture more generally, certainly allow benefits to accrue to some people, they have been roundly critiqued for their environmental, health, and social costs, which are mostly borne by those who do not directly benefit. Pesticide problems are especially severe in market-oriented vegetable and fruit cropping systems, which typically depend heavily on synthetic pesticides. Fresh produces' high water content and its resulting perishability make it more vulnerable to pests and pathogens than other crop types. But pesticide use is also higher on vegetables and fruits because their cosmetic appearance matters greatly for actors along the commodity chain, and they generally have much higher values per acre, thereby making investments in pesticides "worth it" for farmers, since spraying protects a crop that represents a considerable investment. This high level of pesticide use intersects with horticultural production's heavy reliance on hand labor, meaning that exposure levels are often very high because of the common presence of workers in the field (Nash 2006). For these reasons and because vegetable production and markets have been growing rapidly in many areas of the world, pesticide use in vegetable production is the substantive focus of this book.

Political Ecology as a Research Framework

I use a political ecology approach to explain the relationships between market requirements, farm households, ecological conditions, and pesticide use in the study site. Pesticides are an intriguing political ecological puzzle, since their use arises largely through political-economic integration into markets and is influenced by a large range of processes and conditions, including agrifood system governance, unequal power relationships in the commodity chain, farmers' access to resources, crop value, and climate, pest organisms, and crop susceptibility (Galt 2008c). Political ecology offers a particularly useful approach for exploring these intersections.

As an academic field, political ecology remains loosely bound but extremely vigorous (Robbins 2012). A seminal definition by Blaikie and Brookfield (1987, 17) is that political ecology "combines the concerns of ecology and a broadly defined political economy. Together this encompasses the constantly shifting dialectic between society and land-based resources, and also within classes and groups within society itself." Other political ecologists describe the approach as one that combines cultural ecology's focus on behavioral aspects of local resource use and a structuralist political economy, with "rigorous attention given to *production* processes at various scales of analysis" (Bassett 1988, 470, emphasis in original). More broadly, political ecology asserts that "environmental problems cannot be understood in isolation from the political and economic contexts within which they occur" (Bryant and Bailey 1997, 28). Political ecology differs from the field of environmental politics in that it pays considerable attention to the specificities of organisms and ecological interactions and how these shape and are shaped by society rather than seeing them only as a stage on which politics, large and small, plays out (Walker 2005). Political ecologists tell stories of justice and injustice, of winners and losers, and "it is essential to understand the degree to which such outcomes are non-incidental, persistent, and repetitive: a structure of outcomes that produces losers at the expense of winners" (Robbins 2012, 87).

These definitions suggest that political ecology contains considerable diversity in topics analyzed and methods used. Political ecologists have created theoretically informed case studies on a number of topics in human-environment relations, including deforestation and forest protection, conflicts created by parks and protected areas, soil erosion, rangeland degradation, famine, contract farming, export agriculture and local food production, agricultural water use, biotechnology and genetically modified organisms, and agrochemical use, to name a few. Although one cannot precisely map the boundaries of political ecology, below I draw on political ecology's theoretical foci relevant to the book: structural and the governance of commodity chains.

The influence of powerful political economic forces acting from afar on land users' decision making has long been a central concern of political ecology. Much early political ecology was structuralist in orientation, highlighting social structures and their impacts rather than individuals' interpretations of them or their actions within them. A prominent analytical metaphor of political ecology, the "chain of explanation," reflects structural thinking: "It starts with the land managers and their direct relations with the land (crop rotations, fuelwood use, stocking densities, capital investments and so on). Then the next link concerns their relations with each other, other land users, and groups in the wider society who affect them in any way, which in turn determines land management. The state and the world economy constitute the last links in the chain" (Blaikie and Brookfield 1987, 27). Following the "chain of explanation"—extending analysis from the field to the larger-scale political, economic, and ecological actors and processes that affect land users' decisions—has been successful in the study of land and water degradation in developing countries. The analysis in this book shows that it is equally useful to understanding how regulations affect land-use practices and create new spatial arrangements of agrifood governance.

In research on agroexport crops, the last link in the chain of explanation was viewed in a one-sided fashion by those in the political ecology and political economy tradition in the 1990s. As noted above, these critics argued that export markets pressure farmers to increase pesticide use to comply with cosmetic and phytosanitary requirements, which makes export crops more pesticide intensive than local or national market crops (Murray and Hoppin 1992; Thrupp, Bergeron, and Waters 1995; Wright 1990). These accounts are specific expressions of political ecology's well-articulated and much broader "degradation and marginalization thesis" (Blaikie 1985; Blaikie and Brookfield 1987; Robbins 2012). This thesis argues that "otherwise environmentally innocuous production systems undergo transition to overexploitation of natural resources on which they depend as a response to development intervention and/or increased integration in regional and global markets. This may lead to increasing poverty and, cyclically, increased overexploitation. . . . [M]odernist development efforts to improve production systems of local people have led contradictorily to decreased sustainability of local practice and a linked decrease in the equity of resource distribution" (Robbins 2012, 21).

The conceptualization of a structurally determined relationship has been questioned in more recent research into the governance of agrifood commodity chains. Researchers have paid more attention to the governance of agrifood commodity chains by the state and, increasingly, private-sector standards (Barrett et al. 1999; Bingen and Busch 2006).[11] Following Lowe, Marsden, and Whatmore (1994, 2), regulation and standard setting in agrifood systems should be construed broadly as "economic governance" to avoid "a false antithesis between the state and the market." This economic governance, like environmental governance, acts "to assure the stability of capitalist relations of power and accumulation" (Robertson 2004, 362). This newer work, much of it done through

ethnographic methods attentive to the lived realities of various participants in the commodity chain, has shown that food anxieties in the industrial world have strong impacts on developing countries throughout the
commodity chain and especially on the farms where production occurs
(Dolan and Humphrey 2000; Fischer and Benson 2006; Freidberg 2004).

Although previous work told stories of the destruction of local cultures
and ecologies, the common narrative strategy of summarizing these effects as all negative or all positive is untenable once one understands the
complexities of the changes involved (Carter and Barham 1996; Damiani
2003; Fischer and Benson 2006). Fischer and Benson (2006) provide a
nuanced account of new agroexports in highland Guatemala. Summing
it up as "a cheap supply of produce in the United States versus a little
extra cash and a whole lot of risk for Maya farmers" (3), they do not dis-
miss outright the new agroexports and instead tell the story through the
views of those participating. They argue that desire underlies both U.S.
consumers' shift toward more fresh produce and Maya producers' adoption of broccoli and other new agroexports to have "something more."
The account details "how economic inequalities on a global scale can dig
deeply into a local world, not simply through crude forms of extraction
and force but through subtle channels of desiring" (44). About changes
in production, they note that "Maya broccoli farmers need to adopt
production methods that involve new forms of education and training
in order to meet the arguably excessive quality standards of American
consumers" (9). In regard to agrochemicals, this includes adopting new
"forms of knowledge and discipline that were not necessary just years
ago" (30).

A number of studies now show that export-market regulations can
positively impact export farmers' pesticide use in terms of reduction in
intensity or adoption of less toxic chemicals (Arbona 1998; Boh 2003;
Julian, Sullivan, and Sánchez 2000; Okello and Swinton 2007; Okello and
Okello 2010; Opondo 2000; Williamson 2003). These findings and the
food systems governance literature suggest that the common thesis that
integration into export markets causes increased degradation put forth
by those using a political ecology perspective to understand new agroexports needs to be empirically evaluated in light of new developments.

More generally, this recent work has shown that the way that regulation from afar works in local contexts requires paying more attention to local geography and history—and to the understandings held by people at various places in the commodity chain—than agroexport critics have done thus far.

Much of the more recent work notes the contingency of outcomes in new agroexports. Work by Freidberg (2004) on fresh vegetable commodity chains focuses on the cultural context of economic relationships and shows how outcomes of markets depend on different cultures and social networks. Her parting warning for a world increasingly relying on "codified standards of goodness" is that applying them to "people and their goods threatens to put more and more of the food supply in the hands of actors who know everything about category management and nothing about the taste of a good carrot; who know how to audit but not how to intuit. It threatens to erode the social basis of trust in food trade networks, and especially in those networks spanning long distances and varied social and cultural contexts" (222).

This seems to be a warning largely for alternative agrifood commodity chains—fair trade, organic, and, potentially, local—in which codification has become a common strategy to enhance their fairness or sustainability. Establishing standards fits very well with neoliberalism, which privileges market-based strategies and audit culture (Guthman 2007). It is these new alternative agrifood commodity chains and their governance that have received a disproportionate amount of attention from critical scholars over the past decade. Indeed, the focus on alternative food systems has meant that important problems in conventional agriculture that were prominent features of previous critiques—including the effects of industrial agriculture inputs on soil, water, and off-site ecological conditions and how these might be better regulated—have largely been absent from political ecologies of late. It is in these commodity chains rooted in industrial agriculture that abandoning standards would be deeply problematic, since, as we will see, behaviors that endanger consumers and environments can have substantial economic benefits for farmers. Indeed, for better or for worse, we need standards to govern conventional commodity chains in the food system as well as many other aspects of modern existence (Busch 2011).

It is in this context that I develop a political ecological understanding of pesticide use in intensive agricultural systems and of how pesticide governance works in very unequal ways through the food system. A handful of political ecologists have worked on this topic, especially in the process of production. Lori Ann Thrupp, strongly influenced by Blaikie's political ecology, points out that high pesticide use "cannot be seen merely as a 'natural' reaction to high pest incidence" (Thrupp, Bergeron, and Waters 1995, 49). Instead, she points to many other factors—largely structural—that influence pesticide use, including stringent market requirements for cosmetic perfection, national policies that create incentives for pesticide use, agricultural credit policies that often mandate pesticide use as a condition for receiving a loan, active (and often dishonest) sales promotion by agrochemical companies, intermediaries who provide technical assistance to contract farmers and promote and sell agrochemicals, and farmers' inflated perceptions of pest risks (Thrupp 1988, 1990a, 1990b, 1991a, 1991b, 1996). Grossman (1998) expands upon Thrupp's structuralist interpretation of pesticide use. In addition to structure, Grossman emphasizes farmers' agency by positing that farmers' considerable differences in pesticide use can be explained by individuality and propensity to experiment.

While this work has been influential in my own work, I seek to advance the political ecology of pesticides on two main fronts. First, following the recent attention to food systems governance, I devote a great deal of attention to how governance—particularly extraeconomic regulation—functions in real-world markets. Polanyi argued that markets are always embedded in social relations (Block 2003; Polanyi 1957). Social regulation, or in Marxian and Polanyian terms extraeconomic regulation, refers to "those forms of regulatory control that are not directly concerned with the control of markets or other specific aspects of economic life, but instead aim to protect people or the environment from the damaging consequences of industrialization" (Hawkins 1989, 663). Rather than being antagonistic to capital, extraeconomic regulations "nonetheless support and sustain capital accumulation" (Jessop 2001, 215).

Pesticide residue regulation is a prime example of extraeconomic regulation. I pursue the question of how market signals and market governance—concerning both quantity but especially various extraeconomic

qualities that allow for compliance with standards created by regulation—are transmitted, or socially mediated, through the contract farming
relationship that export firms use in coordinating supply chains. Political
ecologists have paid attention to the social relations of exchange within
contract farming, but this has been largely conceptualized in a way that
views contract farmers as unfree (Watts 1992). Contract farming is not
just a formalization or crystallization of exploitation in the social relations of exchange but is also a relationship between actors who, despite
different and often conflicting class interests, generally share the interest
of meeting extraeconomic requirements of their markets. They are therefore involved together in governance, yet different interests and asymmetries in power make these relationships potentially conflictual, either
openly or, more commonly, covertly. Complying with market requirements in the context of unequal power relations makes for a more interesting micropolitics—and attendant ecological dynamics—than seeing
contract farming as just being about exploitation. Thus, this book aims
to expand our understanding of food systems governance by comparing
different market channels, export and domestic.

Second, I aim to show the importance of the biophysical environment's
influence on pesticide use and social change. Many political ecologists
reject the ontological divide between nature and society, showing how
human and nonhuman actors together create nature-culture, socionature,
or nature-society hybrids or fusions (Castree and Braun 1998; Swyngedouw 1999; Whatmore 2002; Zimmerer and Bassett 2003). While Grossman (1998) argues for the importance of the "environmental rootedness"
of agriculture, he only cursorily shows how it shapes pesticide use and
its consequences. Robbins (2012) argues for increased recognition of the
agency of nonhuman actors in human-environment interactions. The activities and life cycles of insect pests and bacterial, fungal, and fungi-like[12]
diseases are a precondition for pesticide use, but they are not just a backdrop because they are living organisms, having all the unpredictability
that organisms and their populations bring. These pest populations both
shape and are shaped by pesticide-use regimens. Since climate strongly
influences the distribution and prevalence of these species, climate and
all of its various components should be considered an explanatory factor
in pesticide use. Thus, I aim to show how biophysical processes together

with political economy and its governance jointly shape the dynamics of agrifood systems. A political ecology of agriculture must recognize both political economic processes and environmental processes as well as their entanglements and mutual shaping of one another.

Plan of the Book

The structure of the book reflects the comparative case study approach that I used in the research. Chapters 1, 2, and 3 focus largely on the dynamics of agrarian capitalism and market integration, socioeconomic differentiation, environmental conditions, and pesticide use in the study site. Chapters 4 and 5 focus on new agroexport production and its shaping from afar by both capital and the state. Having established the dynamics of both sectors, I present in chapter 6 detailed comparisons of the dynamics of agricultural production and pesticide use for the national and export markets.

Chapter 1 introduces the social, agricultural, economic, and environmental conditions of the study site. The last part of the chapter puts pesticide use in a political economic context by connecting it to important concepts, including industrial appropriation, circuits of capital and capital's spatial fix, the contradictions of capitalism, and the treadmill of production. This understanding of agrarian capitalism serves as the basis for the ensuing chapters.

Chapter 2 examines the local geography of pesticide use by looking at the spatial variation in farmers' pesticide use, especially as it relates to climatic heterogeneity and to farmers' resources. Analysis focuses on the cloud belt, an area more commonly in the clouds than not, that promotes intense disease pressure. Wealthier farmers who live in the cloud belt are more likely to access better growing environments outside the cloud belt for their crops, thereby lowering their pesticide use and producing higher crop yields and higher profit rates. Poorer farmers who live in and farm only in the cloud belt remain trapped in poor climates for their crops and spray more heavily. Socioeconomic differentiation is mediated through unequal access to a heterogeneous environment and shows how political ecology's degradation and marginalization thesis applies to intensive industrial agriculture.

Chapter 3 provides an agroecological history of the study site, focusing on landscape change, annual crops, market integration, and agrochemical use. Potato production for the national market began in 1910 and initially did well without agrochemical inputs, but by the 1950s, potatoes faced devastating imported pests. To save their crops and increase their yields, farmers turned to synthetic fungicides, insecticides, and fertilizers. By the 1960s, farmers in the area were the most likely to use agrochemical inputs in all of Costa Rica. By the 1970s, vegetables produced for the national market relied heavily on agrochemicals. It was into this context of high synthetic pesticide and fertilizer use that the new agroexports of the 1970s and 1980s were introduced. In the 1980s, the new agroexports faced rejections from the U.S. market due to illegal pesticide residues, a situation to which the state and agroexport firms responded.

Chapter 4 focuses on agroexport firms as supply chain coordinators, including their responses to pesticide violations, understandings of export market regulations, and interactions with their contract farmers. Export firm managers have strong concerns over rejections due to illegal pesticide residues. Their attempts to control farmers' pesticide use began shortly after the first rejections due to illegal pesticide residues found by U.S. enforcement. Their current understandings of export market regulations are shaped by this history, and exporters do not completely understand U.S. pesticide residue regulations. Yet the forms of control they have established are largely working, although they are certainly not perfect. This means that capital's control over farmers' labor process cannot be predicted with certainty, as local context matters greatly.

Chapter 5 turns to export farmers, connecting recent discussions of the impacts of the industrialized world's agrifood regulation on farmers in the global South with the literature on pesticide use in developing countries. Previous attempts to explain how agrifood regulation impacts farmers' pesticide use are lacking because they do not situate it deeply enough within the social relations of exchange in contract farming. Many farmers exercise considerable caution in their pesticide use vis-à-vis residues on the main export crops from the study site. Exporters' mediation of regulatory risk—conceptualized as the possibility that an actor's behavior will be subject to state regulation and that out-of-compliance behavior will result in negative consequences that impact the actor—largely explains this type of caution. A gap between regulation and practice persists, however, because of the local history of residue violations and a related misinterpretation of pesticides' label color bands.

Chapter 6 addresses pesticide residues on vegetables in developing countries through the Costa Rican case. Pesticide residues are often very high on vegetables in developing countries, generally considerably higher than in industrialized countries. I examine why vegetable farmers use pesticides in such a way that high levels of residues remain and how markets with unequal regulatory strength affect farmers' pesticides use and the resulting exposure of different populations fed by different market segments. While usually attributed to farmer ignorance, the pesticide residue problem arises from a triad of causes: higher efficacy of more residual and toxic pesticides, the biological characteristics of vegetables that mature continuously and require harvests every few days, and a volatile national vegetable market upon which farm households depend. With high input costs and low farm-gate prices, farmers in markets with minimal regulation tend to use more residual and toxic pesticides. Additionally, with lax regulation in the open national market, some export farmers use it as an outlet for their produce when they use highly residual pesticides.

The conclusion points out that pesticide use does not have to be so high in the area and develops implications of the study. I note the insights into industrial agriculture provided by a political ecological approach and then lay out detailed recommendations for reducing pesticide use in the area and in Costa Rica as a whole.

1 Farm Households, Environmental Geography, and Agrarian Capitalism

In October 2003 while I was conducting my fieldwork, Costa Rican news outlets reported a disturbing event: elementary schoolchildren in El Yas, a small town a few kilometers south of where I was living, were poisoned by methamidophos—an acutely toxic organophosphate insecticide—drifting into their classrooms. The school in El Yas is located immediately next to farm fields, as is the case in many other towns in this agricultural area. In the middle of a lesson, children began feeling dizzy and vomiting. Five children were taken to the hospital, which fortunately is only ten kilometers away via decent roads. A neighboring farmer growing chayote, a cucurbit native to the region with a pear-shaped fruit and grown on a trellis, had applied the pesticide, which is prohibited by exporters for the export market, on his fields earlier in the morning. A breeze brought it through the classrooms, exposing the children (Gutiérrez C. 2003). Another pesticide poisoning of sixty-one schoolchildren, again involving methamidophos, occurred a few months later nearby in Cartago, although this time it was spraying of the school grounds itself that was the cause (Rojas 2004).

The immediate proximity of heavily sprayed fields and classrooms might be identified as the cause of these events. Yet there are larger-scale processes and contexts that cause these pesticide drift accidents and have contributed to Costa Rica's heavy pesticide dependence. Indeed, it is only through examining these large-scale processes and the way that they unfold in particular areas that root causes of pesticide problems can be elucidated. Failing to do so—that is, "framing of pesticide drift as rare, isolated 'accidents'"—"pushes the scale at which the issue is perceived down to that of the individual incident" (Harrison 2006, 519). Thus, this chapter introduces the farming systems and farming units of Northern Cartago and the Ujarrás Valley and then documents the broader contexts and processes that have led to the region's extremely high pesticide dependence, including environmental conditions and pests. The last section introduces pesticide use as a consequence of the spread of agrarian capitalism, which establishes a number of socioecological contradictions.

Farms and Farming in Northern Cartago and the Ujarrás Valley

The regions of Northern Cartago and the Ujarrás Valley are Costa Rica's "vegetable basket" (figure 1.1). Truck farmers in the region use a range of microclimates and fertile volcanic and alluvial soils to produce more than thirty tropical and temperate vegetables for national and export markets. Commercial vegetable production began about a century ago and now dominates the area's agriculture, together with dairy production in the highlands and coffee and sugarcane in the lowlands.

Farm activities in the area are strongly shaped by political and economic processes operating on a variety of scales. Macroeconomic structural adjustment in the 1980s created reduced state support for nationally oriented agriculture and a stronger export orientation (Clark 1995; Edelman 1999; Paus 1988). National market vegetable production remains strong in the area, however, in part because of tariff protections on potatoes and onions that Costa Rica has maintained even through successive rounds of liberalization and trade agreements (Foreign Agricultural Service 2010). Integration into export markets over the last few decades has resulted in contract farming relationships in vegetable production (Breslin 1996; Mannon 2005; Van Orman 1995), although spot markets are more common for national market produce. Agrochemical formulators and retailers have also aggressively expanded their markets through advertising, state promotion, and private banks requiring agrochemicals as a condition of receiving credit (Farah 1994; Thrupp 1988, 1990b).

Land reform in Northern Cartago in prior decades has made parcels available to smaller-scale farmers (Edelman 1989; Seligson 1980), yet inequality in landholdings remains high. In his geographic analysis of landholdings in Northern Cartago, Arrieta Chavarría (1984, xi) determined that "less than 5 percent of the farms in Cot-Irazú (fewer than 50 farms) are *latifundios* or modern businesses and they have 72 percent of the land (some 11,000 hectares) while 850 farms of the 95 percent possess

Figure 1.1 Towns and physical features of Northern Cartago and the Ujarrás Valley.

some 3,000 hectares, or 28 percent." Most of the large landholdings are dominated by pasture for dairy production, but these farmers also engage in potato and other vegetable production. A few very large vegetable-oriented family-run farming operations farm more than 100 hectares.

The thousands of farm households in the area, which usually consist of nuclear families and often more extended kin, are the social units that bring together land, labor, and capital in intensive vegetable agriculture. Transnational corporate farming operations are largely absent in the area except in fern production (Mo 2001), but a handful of large family-run corporations make land-use decisions in Northern Cartago and the Ujarrás Valley, as noted above. Farmers of the area note a strong individualism that precludes much widespread cooperation, yet farming together with friends and family members is very common as a strategy to meet the high capital requirements of intensive vegetable production.

Farm households in the area are heavily engaged in production for market. Out of 148 farmers and farm managers in my survey (see appendix 1), none are strictly subsistence producers, and only a few grow any crops purely for family consumption. All depend heavily on vegetable sales for family income and fundamentally depend on purchased off-farm inputs, especially synthetic pesticides and fertilizers. Thus, the agricultural economy is not a classical peasant economy of high labor and low capital inputs but instead is highly labor- and capital-intensive. Large expenditures are required to participate in this agricultural economy, and the possibilities for large amounts of capital accumulation, as well as capital loss and thus farm loss, are considerable. For these reasons, I use the word "farmer" rather than "peasant" to refer to agricultural producers in the area because of the connotation with "peasant" of some subsistence production and because the farmers in the area refer to themselves as *agricultores* ("farmers") rather than *campesinos* ("peasants").

The characteristics of these capital-intensive small-scale family farms are shown in table 1.1. Sixty-one percent of households in the survey rely on the labor of more than one family member, and most farms employ only one permanent worker (the median number of permanent workers is one, but large-scale farms in the farmer survey skew the mean for many characteristics; the median is a more reliable measure of the typical situation in these instances). The median amount of land owned is 2.1 hectares, and the median amount of land planted is 2.8 hectares, demonstrating the importance of renting and *a medias* arrangements. *A medias* refers to many arrangements between two or more farmers, including sharecropping (in which a farmer who does not own the land obtains access to it by giving a percentage of the harvest to the land owner or by providing the inputs), family members who farm together but typically

Table 1.1   Characteristics of farm households, Northern Cartago and the Ujarrás Valley, 2003–2004

	Median	Mean	St. Dev.	n
Farmer characteristics				
Age	40	42.1	11.1	148
Years of formal schooling	6	6.5	2.9	148
Years in farming	23.5	23.7	12.3	148
Member of a farmer organization	—	36%	0.5	148
Household & labor characteristics				
Number of minors in household	2	2.14	1.6	148
Number of adults in household	2	3.12	1.5	148
Number of permanent workers	1	4.2	13.6	148
Greatest number of temporary workers at one time	3	5.0	9.2	148
Land ownership & land use characteristics				
Hectares of land owned	2.1	8.4	27.8	147
Current number of hectares planted	2.8	7.0	22.6	148
Number of parcels planted	2	2.1	1.7	148
Number of crops planted	3	3.9	2.3	148
Produces export crops	—	26%	0.4	148
Farm equipment				
Owns a pickup truck or car	—	75%	0.4	148
Owns a tractor	—	16%	0.4	148
Economic characteristics				
Home ownership	—	89%	0.3	143
Value of house(s) owned[a]	$15,050	$20,926	$21,762	133
Received credit in the past 12 months	—	52%	0.5	148
Total agricultural expenses in 2002	$6,948	$17,255	$47,355	133
Total reported agricultural profits in 2002[b]	$2,779	$8,540	$40,778	123

[a] Using the 2003 exchange rate of 398.66 colones/US$.
[b] Using the 2002 exchange rate of 359.82 colones/US$.
Source: Author's farmer survey, 2003–2004.

maintain separate households, and two farmers who farm together and split input costs and profits/losses from sales.

Three-quarters of farmers own a vehicle, although only 16 percent own a tractor and 7 percent own a greenhouse, which can raise yields and reduce pesticide spraying. Most farmers own a house or live in a house owned by someone in the household. Livestock ownership is uncommon by small-scale vegetable farmers the averages. About one-third of farm households receive income from outside of agriculture, which is typically provided by a spouse's or adult child's employment. Slightly more than half of farmers received credit for agricultural production, another indication of its capital intensity (see table 1.1).

The importance of capital in the vegetable production systems of the area is well demonstrated by the crucial role played by pesticides. My survey and all past surveys in the area (Arauz C., Carazo R., and Mora A. 1983; Arnáez et al. 1993; Zúñiga et al. 1991) reveal that essentially all farmers use synthetic pesticides, a situation common in Costa Rica (Hilje et al. 1987). Farmers and farmworkers generally use backpack sprayers, mostly hand-pump backpack models, although motorized backpack sprayers are fairly common. Pesticide application is a highly gendered activity in the area, as is agriculture generally. Male farmers make up the vast majority of the farmers in the area, and all of the farmers in the survey are men. People in the area consider pesticide applications dangerous, and most men do not allow their wives to participate in the activity. During my fieldwork I saw dozens if not hundreds of pesticide applications and never witnessed women spraying pesticides, but a few farmers did mention women's involvement in applications. A norm against young children's involvement in the pesticide application also exists,[1] although teenage boys commonly spray, since many leave school after grade six to work as farmworkers. Farmers clearly know that pesticides are dangerous and do not want their families to run the risk of poisoning. They reserve that risk for themselves, but this does not mean that they wear a great deal of protective gear even though they recognize the hazards (Galt 2013). The region, with the highest documented pesticide use in the world, has extremely high stomach cancer rates (Ortíz Gutiérrez 1996), which most farmers and residents suspect is caused by agrochemicals.

Farmers in Northern Cartago and the Ujarrás Valley grow about three dozen vegetable crops, including artichoke, beet, broccoli, cabbage, carrot, cassava, cauliflower, celery, chayote (figure 1.2), Chinese cabbage, cilantro, corn (local criollo and hybrid sweet varieties), eggplant, endive, fennel, green bean, green onion, jicama, kohlrabi, lettuce, onion, pea, peppers (hot and sweet varieties), potato, radicchio, romanesco, spinach, squash, tacaco, tomatillo, and tomato. The crops that I focus

Figure 1.2   The chayote plant on a trellis. Chayote trellises dominate the landscape in the background. (Source: author)

on most in this book are those grown for both the export and national
markets—carrot, chayote, corn, green bean, and squash—and those that
are most commonly grown for the national market—broccoli, cabbage,
cauliflower, onion, potato, and tomato—although details on other crops
were also collected (see appendix 1).

Different crops dominate different elevations of the area, although
crop distributions are more like overlapping patchworks by elevation

rather than distinct, exclusive crop zones (cf. Zimmerer 1999). The lower elevations in the Ujarrás Valley support fields of heat-loving crops: chayote, cassava, corn, eggplant, green bean, peppers, plantain, squash, tacaco, tomatillo, and tomato. The highest elevations support cold-climate crops: artichoke, carrot, cauliflower, onion, and, most important, potato (figure 1.3). The middle elevations are a diverse mix of many of the warmer-zone crops and the cooler-zone crops and in-between crops such as beet, broccoli, cabbage, cilantro, green onion, lettuce, radicchio, and spinach.

The national market remains the most important outlet for vegetable production in Northern Cartago and the Ujarrás Valley. Potato, the most important national market crop in the region, is planted on 3,000 hectares in Northern Cartago (Centro Internacional de la Papa n.d.; Ramírez Aguilar 1994). While produced almost exclusively for the national market, Costa Rican potatoes are infrequently exported to Nicaragua, although these exports do not have to meet the same standards as exports to the United States, Canada, or the European Union. Potatoes are a very important food item in Costa Rica (see table 0.3), and Northern Cartago produces more than 90 percent of Costa Rica's potatoes, while the next largest production area, Alajuela, accounts for 5.6 percent of production (Centro Internacional de la Papa n.d.). Potatoes are

> a high-input, high-output, high-risk crop. The great responsiveness of yields to inputs—such as high quality [seed] tubers, fertilizers, pesticides, additional labor, and other forms of energy—motivates farmers to use inputs more heavily on potatoes than on other crops. Because of the relatively high level of potato yields, the short growing period, and the high market value of potato tubers, the potato crop generates larger returns per hectare and per day than most other crops grown in developing countries. The susceptibility of the potato to pests, diseases, moisture stress, and extremes in weather makes its yields more variable than those of many other crops. Yield variation, coupled with price fluctuations and high input costs, makes potato production risky. (Horton 1987, 48)

There are dozens of other national market crops produced in the region, the most important of which—carrot, cabbage, and onion—are grown on many hundreds of hectares.[2]

Figure 1.3   A potato harvest on one of the parcels of Finca Guarumos, high on the eastern side of Northern Cartago. (Source: author)

Two vegetable agroexport sectors also exist in the area: chayote and minivegetables. Farmers started growing chayote for export in the mid-1970s (Bolaños et al. 1993). Of the 555 hectares of chayote in the Ujarrás Valley, half are devoted to export and half to the national market (Secretaría Ejecutiva de Planificación Sectorial Agropecuaria 2004, 51). The minivegetables—including mini–scallop squash, minizucchini, mini-carrots, fine green beans, and others—were introduced for exportation in the mid-1980s (Breslin 1996) as part of the export-oriented "Agriculture of Change" that the Costa Rican state promoted as part of structural adjustment imposed by the International Monetary Fund in the early and mid-1980s (Edelman 1999; Gaete 1990). Minivegetables started appearing in French nouvelle cuisine in the 1970s and 1980s and can retail for $7 per pound (Freidberg 2009, 193). The export sector orients minivegetables to the higher-paying export market, although some are sold nationally as well. The portion sold nationally is not determined by farmers, however, since the minivegetable exporters buy all farmers' minivegetables and decide the level of quality that goes to the different markets at different times of year.

The vast majority of the exports of chayote and minivegetables go to the United States and Canada, although a small proportion of chayote goes to Europe. The export variety of chayote, *quelite*, which has a pale green and smooth skin, is also sold openly on the national market. Unlike minivegetable farmers, export chayote farmers are not obligated to sell all of their *quelite* to the exporter, nor do they (Mannon 2005). Farmers grow the other chayote varieties almost exclusively for the national market.

Production systems aimed at the different markets—national and export—exist side by side in the study site, often cultivated by the same farmer, making it an ideal location to comparatively understand their dynamics vis-à-vis ecological and political economic processes and contexts. The next section explains the environments in which these crops are grown and what it means for the various disease and insect pests that farmers' crops face.

Environmental and Pest Geography

During my fieldwork I would often travel many kilometers away from my midelevation home in Cipreses, heading down into the hot Ujarrás Valley or up into the cold highlands of Volcán Irazú. Cipreses, at seventeen hundred meters above sea level (masl), or about fifty-six hundred feet, was in the clouds more often than not. As I traveled, I noticed that this cloud belt generally did not extend many kilometers north or south. During interviews and informal discussions, many farmers, especially those from Cipreses who planted faraway fields, noted that the Cipreses area and similar elevations to the east of the volcano are indeed cloudier, guaranteeing very high levels of humidity and favoring plant disease. The first interview where this spatial variation came up was toward the beginning of my fieldwork, when I was talking with Nelson, a large-scale minivegetable grower. Nelson noted that he refuses to plant minivegetables in Cipreses: "Here we have a problem with production. One has to use lots and lots and lots of pesticides."

To understand the boundaries of the area, I asked, "Where does the zone end, more or less?"

"From the cross [at the fork in the main road southwest of Cot near Cuesta La Chinchilla], toward here; it is tremendous; it [experiences] what is called a *temporal;* it is not a storm that comes with a downpour, passes by, and the sun comes out. No, it is gloomy, gloomy days that have very little light. This affects [production]."

I followed with, "And above, up to where?"

"No, it is not above. From Cot and above, there is more sunlight."

Still trying to pin down the zone's boundaries, I asked, "And below, down to where?"

"Below, until Cervantes . . . there is not much light. It is more like toward the center of Cartago, from El Guarco [Valley]." Through this dialogue, Nelson revealed his mental map of the cloud belt, the area often enveloped in clouds and strongly affected by *temporales,* which are intense multiple-day storms during which the sun does not shine.

My research in Costa Rican libraries months later led me to find a map of sunlight in Costa Rica, which I superimposed on the map of topography and towns of the area (figure 1.4). It is striking how well Nelson's mental map, my own experiences, and the climatological map correspond. On figure 1.4, the dark area inside the "4 hours" line experiences, on average, fewer than four hours of direct sunlight per day. This is the approximate extent of the cloud belt. The lighter area bounded by the "5 hours" line experiences more than five hours, while the area between receives from four to five hours of direct sunlight per day.

The cloud belt forms due to the dominant easterly trade winds that bring cloud cover and precipitation from the east. The moisture-laden air from the Caribbean hits the eastern slopes of Volcán Turrialba (to the northeast of the map) and Volcán Irazú and rises due to orographic lifting. This forms dense clouds as the moisture in the cooling air condenses, and the clouds are routed to the southwest along the flanks of the volcanoes. The cloud belt covers this midelevation area for much of the rainy season and is less pronounced but often still present in the dry season. While it rains very often in the region, the cloud belt itself does not always produce rain but always brings very high levels of humidity.

The dramatic topography of Northern Cartago and the Ujarrás Valley creates other strong environmental gradients. Average annual temperature decreases with elevation, ranging from more than 20°C (68°F) in Ujarrás

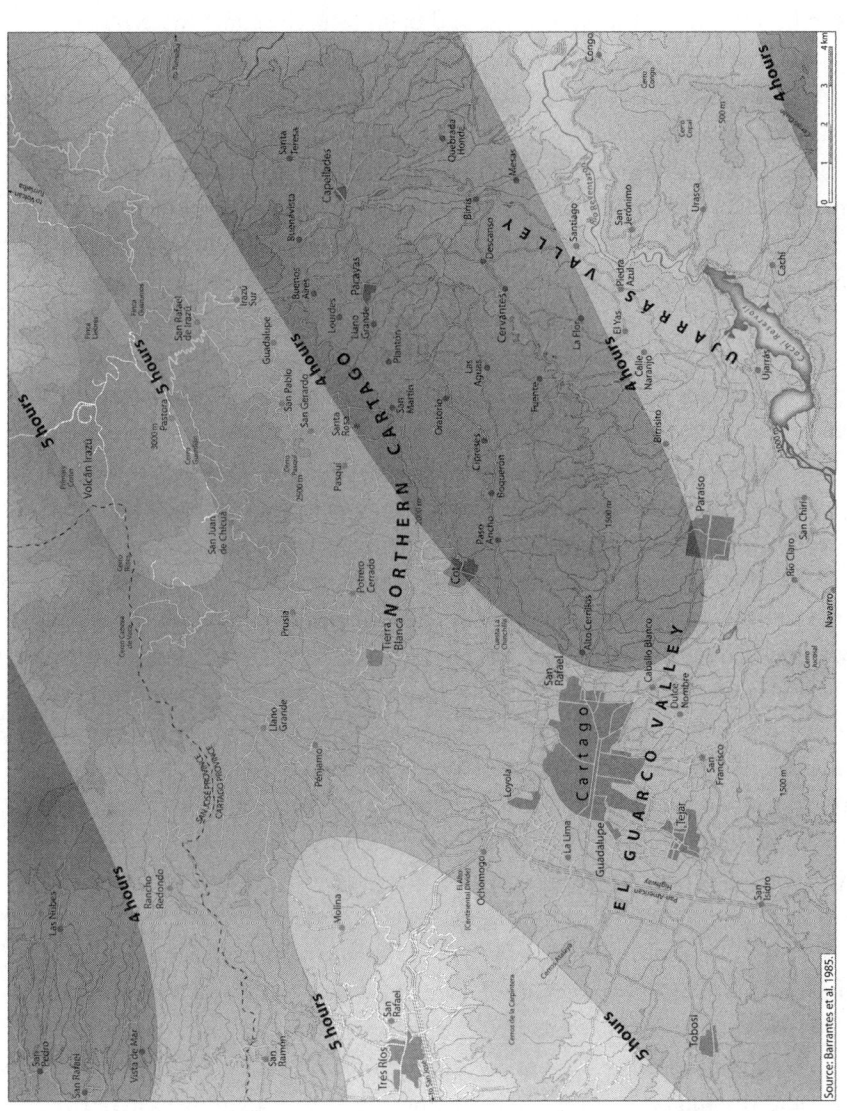

Figure 1.4 The cloud belt as shown by daily hours of sunshine in Northern Cartago and the Ujarrás Valley, annual average from 1961 to 1980.

to about 10°C (50°F) in San Juan de Chicuá (Barrantes F., Liao, and Rosales 1985). Annual average precipitation does not correspond directly to
cloud patterns, instead following a strong east-west gradient with a slight
bulge in higher precipitation in the cloud belt along the southeastern flank
of Volcán Irazú. Capellades in the east receives 2,500 mm (8.2 feet) of
rain annually, while Cartago in the west receives less than 1,400 mm (4.6
feet). This large range in annual precipitation is caused by the volcanoes,
with orographic lifting on the eastern slope and a rain shadow on the
western slope. The rain shadow is strong; the amount of rain in the area
west of Cartago is on par with the driest part of the country, the Nicoya
Peninsula on the west coast (Barrantes F., Liao, and Rosales 1985).

The rainfall and often thick cloud cover of the area used to support
cloud forest and, to a lesser extent, rain forest vegetation. Life zone maps,
based on Holdridge's influential climate-based vegetation classification
system, show rain forest, wet forest, and moist forest covering the area
(Tosi 1969). The last two categories dominate almost all areas of the
study site and have similar meanings to the common understanding of
the term "cloud forest" supported by very high humidity levels. While
almost all of the forest has been removed for agriculture, remnant trees
and patches of forest show common cloud forest characteristics, especially strong coverage by epiphytes.

These environmental conditions—humidity, temperature, and rainfall—strongly influence agriculture in the area, since they are important
ecological conditions for insect and disease pests. Vegetable farmers in
the area face very high levels of insect pest and disease pressures, as suggested by the extremely high levels of pesticide use. The following para-
graphs serve as a primer on disease and insect pests to which farmers are
responding. Since there are dozens of pests and diseases that affect the
crops of the area, the treatment here is very selective, with a focus mostly
on potato and then on squash. Table 1.2 shows a number of squash and
potato diseases and relates them to the environmental geography of the
area. Late blight—*Phytophthora infestans* and known locally as *quema*
("burn") and *mancha* ("stain")—is the most potent fungal disease in the
area, so I describe it first.

Late blight affects potatoes and tomatoes. It is the most devastating
plant disease in the history of humanity (Abad and Abad 1995). Late

Table 1.2 Environmental influences on important pathogens of potato and squash, Northern Cartago

Squash

Local Name	English Name	Species	Environmental Influences	Source
mildiu	downy mildew	*Peronospora parasitica*	Promoted by high humidity and heavy dew, fog, and frequent rain	Bernhardt, Dodson, and Watterson (1988), Kuepper (2003)
quema/mancha	squash phytophthora blight	*Phytophthora capsici*	Infection and spread favored by excess moisture and standing water	Hausbeck and Lamour (2004)

Potato

Local Name	English Name	Species	Environmental Influences	Source
quema/tizón tardío	late blight	*Phytophthora infestans*	Sporulation requires eight or more hours of leaf wetness; optimal conditions are high humidity, high rainfall, and moderate temperatures (10–20°C)	Apple and Fry (1983); Haverkort and Bicamumpaka (1986) and Devaux and Haverkort (1987), cited in Haverkort (1990); Torres (2002)
maya	bacterial wilt	*Pseudomonas solanacearum*	Not found locally above 2,220–2,500m; very problematic below 1,500m	Jackson (1983), Sáenz Maroto (1955)
pierna negra	blackleg	*Erwinia carotovora*	Problematic above 2,500m	Jackson (1983)
mosaico rugoso	potato virus	20+ viruses (some spread by insect vectors, others by seed tubers or tools)	Insect vector (the aphid *Myzus persicae*) less prevalent at elevations above 2,500m	Hord and Rivera (1998), Meneses (1990)

blight has devastating consequences, with the ability to wipe out entire fields under the right conditions. It is infamous for its role in the Irish potato famine of the mid-nineteenth century. The severity of late blight attacks can make fungicide use on potatoes very high, especially in areas conducive to the disease: "To prevent the buildup of lesions that serve as sources of infection, the fungicide must be present on the foliage at the time of inoculation [when fungal spores land on the plant]. Farmers often begin spraying early in the growing season before the first attack is expected. . . . Where blight occurs, and where fungicides are available, farmers usually spray every 3 to 20 days, depending on the probability and severity of the attack" (Horton 1987, 41–42).

Late blight is a water mold, or oomycete. For reproduction, it depends on leaf wetness for periods of at least eight to ten hours (Apple and Fry 1983; Hijmans, Forbes, and Walker 2000). Late blight reproduces sexually and asexually. Sexual reproduction occurs between opposite mating types (A1 and A2) and produces resistant oospores that survive without a host. A1 and A2 types have coexisted historically in Mexico, while most of the world until recently was plagued only by the A1 type (Zwankhuizen, Govers, and Zadoks 1998). Tests of mating types in Costa Rica reveal that all are A1, so only asexual reproduction occurs (Sánchez Garita, Shattock, and Bustamante 2000), requiring a continuous host. This need is met in Northern Cartago and the Ujarrás Valley, since tomato and potato—its two important host species—are planted year-round, and native *Solanaceae* persist in field edges. These conditions allow the fungus to maintain a high store of spores for inoculation that are easily spread by wind, rain splashes, animals, and human activity.

Once the spores land on the potato leaves, they delve into it with threadlike filaments called hyphae. The mass of hyphae, the mycelium, grows inside of the leaves (Stern 1997). Lesions show after three to five days in the United States (Apple and Fry 1983), but in this amount of time infection may spread so rapidly in Costa Rica as to destroy a field (Molina Umaña 1961, 9). Each lesion produces thousands of spores, which are then spread again by wind and rain to healthy plants, restarting the cycle. Infected plants lose leaves and stems, and valuable tubers can rot in the soil or after harvest.

The optimal conditions for late blight are high humidity, high rainfall, and moderate temperatures (Haverkort and Bicamumpaka 1986 and Devaux and Haverkort 1987, cited in Haverkort 1990, 258; Torres 2002, 5). "Certain specific environmental conditions are necessary for the abundant production and dissemination of sporangia: nights with relative humidity between 90 and 100 percent and a temperature between 10 and 20°C [50–68°F], and also days with some rain and constantly cloudy skies. When these conditions are repeated for several successive days, it is certain that the production and dissemination of sporangia (easily carried by the air), as well as the infection of healthy tissue, will be very high and an outbreak can be predicted" (González 1975, 2). Late blight development during clear, sunny, and dry days "is stopped completely, but on cloudy, humid, and cool days it grows rapidly, covering a leaf in a few days" (González 1975, 1).

In areas where farmers do not depend on fungicides for late blight control, they plant potatoes seasonally to avoid severe late blight problems in the rainy season. In contrast, in areas where planting occurs throughout the year, such as the Kenyan highlands (Haverkort 1990) and Northern Cartago, farmers depend heavily and consistently on agrochemical inputs, especially during the rainy season when relative humidity is especially high.

Northern Cartago has perhaps the greatest late blight pressure in the world. Writing about Northern Cartago, Molina Umaña (1961, 9) notes that with "strong outbreaks of *P. infestans* . . . a field can be eliminated in a period of one to four days." Climatic and cropping conditions on Volcán Irazú are ideal for the sporulation of the fungus, so "weekly fungicide spraying is necessary" (Jackson 1983, 104). The Centro Internacional de la Papa (International Potato Center), known by its Spanish acronym CIP, has conducted surveys that indicate that Costa Rican potato farmers use the largest number of fungicide applications, fifteen per crop cycle, of all nations surveyed. Next highest are the Dominican Republic and Cuba at twelve, while Guatemala and Nicaragua have an average of six and three fungicide applications per growing cycle, respectively (Hijmans, Forbes, and Walker 2000, 704).

Because of late blight, potato farmers in Northern Cartago depend on the constant introduction of new late blight–resistant potato varieties by the state. Resistance, however, is sometimes only effective for a short period because of the very high disease pressure. Since the 1950s, many resistant varieties have come from Mexico (Horton 1987, 42). An early Costa Rican potato manual (Sáenz Maroto 1955, 20) notes that many of these "do not have resistance to late blight, even when they were very resistant in their original environment." For example, in the 1980s, the Atzimba variety was bred for late blight resistance in Mexico, but in Costa Rica, the combination of ideal climatic conditions for late blight and the widespread planting of Atzimba "contributed to the selection of physiological races of the pathogen that overcame the genetic resistance of the variety" (Jackson 1983, 104). Thus, potato farmers in Northern Cartago struggle daily against late blight to see their potato fields through to harvest, and their primary weapons are a range of fungicides manufactured by transnational agrochemical companies and Costa Rican formulating businesses.[3]

Major potato insect pests include two species of tuber moth. The females lay their eggs on the underside of leaves and exposed tubers. Tuber moth larvae bore through tubers, making them unmarketable (Jackson 1983). Farmers combat them with granulated insecticides applied to the soil, which are the norm in potato production. In 1989 a leaf miner, *Liriomyza huidobrensis,* became an important secondary pest in the area, greatly reducing potato yields that year since farmers had no good control methods. This leaf miner continues to cause problems in potato and many other horticultural crops (Comité Técnico de *Liriomyza* 1990; Rodríguez 1997). Most insect pests cause greater problems at lower elevations in the area, and leaf miners only cause major damage to potato at elevations under twenty-four hundred masl (Rodríguez 1997, 3).

Squash is another common vegetable crop in the area. Unlike potato, which is for the Costa Rican market, squash is grown for both export and national markets. The relationships between squash and its diseases and the environmental conditions of the area are similar to potato. The diseases downy mildew (*Peronospora parasitica*) and squash phytophthora blight (*Phytophthora capsici*) are the most important pathogens affecting minisquash in Northern Cartago (HortaRica agronomist, interview, December 2003). As with late blight, the cloud belt environment in Northern Cartago is ideal for these diseases. Downy mildew is promoted by high relative humidity (Bernhardt, Dodson, and Watterson 1988) since it "is most aggressive when heavy dews, fog, and frequent rains occur" (Kuepper 2003, 1). *P. capsici,* the same genus as potato late blight, can lead to the loss of an entire crop (Wasilwa, Correll, and Morelock 1995, 1188), and "excess moisture is the single most important component to the initial infection and subsequent spread of *P. capsici*" (Hausbeck and Lamour 2004, 1299). Even planting in well-drained fields cannot necessarily avoid disease development: if a rainfall event drops more than 2.5 centimeters, standing water may remain long enough for the release of zoospores (Hausbeck and Lamour 2004, 1299). Squash farmers, like potato farmers, deal with these diseases through frequent fungicide applications.

The ecological gradients in Northern Cartago and the Ujarrás Valley strongly affect these disease and insect pests. Like Nelson, who refuses to grow his minisquash in the cloud belt, farmers who grow potatoes in different areas recognize the influence of these agroenvironmental gradients. I traveled with some farmers who lived in the cloud belt up to the parcels they plant above it to ask about geographic differences in pests. Dagoberto, from Cipreses, elevation seventeen hundred masl, described the situation while we were on his rented field in San Juan de Chicuá, elevation twenty-seven hundred masl.

DAGOBERTO: Potato below [in Cipreses] lasts three and a half months; then you harvest. Here [in San Juan de Chicuá] it lasts until four months, four and a half months, and you harvest. There are always about twenty-two days more here than below. But the quality of [potato] is better. . . . Here there is not as much bacteria, not as much fungi, nor as many problems as below. . . . It is sprayed in longer intervals; it is healthier. [In the dry season, every] eight days spraying below, here it is fifteen days, every fifteen, depending on the season. . . . Below potato is sprayed two times [per week] in the rainy season. . . . Here, one time [per week], nothing more. . . .

RG: But is the pressure of bacteria and fungi much less than below?

DAGOBERTO: It is less, yes, sure. Less fungi and less bacteria here than below.

RG: But insects?

DAGOBERTO: There are plenty of insects, but not as many as below. Because above, there are many that are not present. (Farmer interview, April 2003)

I also conducted participatory mapping activities with seven farmers who plant potato in more than one location (see Appendix 1). All seven drew an elevational pattern for late blight, agreeing that higher elevations have lower late blight pressure.

Farmers in the area, then, face diseases and insects that threaten their crops. Since they rely on those crops for their livelihoods, farmers turn to pesticides, which input suppliers are more than happy to provide. Thus, farming in the area is heavily shaped *by* and *for* capital.

Industrial Agriculture and Pesticides: A Faustian Bargain

Before the advent of agrochemicals, farmers generally had to grow crops that were well adapted to a local environment, including its insect pests and diseases. Sir Albert Howard, a pioneer of the organic agricultural movement, noted that pests and diseases were "nature's censors" that farmers could use for "pointing out unsuitable varieties and methods of farming inappropriate to the locality" (Howard 1943, cited in Pollan 2006, 150). Howard's view of agricultural production applied to Northern Cartago and the Ujarrás Valley would mean that most temperate and highland tropical vegetables should not be grown in former cloud forest environments, since they are not well adapted to the area. But this ecologically oriented view is antithetical to the development of agrarian capitalism, in which the circulation of capital for further accumulation, not ecological compatibility and well-being, is the underlying law of motion.

Agricultural industrialization depends on new crop varieties, usually more responsive to synthetic fertilizers, and pesticides as a form of protection. As farmers adopt synthetic fertilizers, they tend to decrease their use of manure and leguminous green manures that fix nitrogen from the atmosphere. This shift affects plant and soil qualities, with important ramifications for pests and pesticide use: "the ability of a crop plant to resist or tolerate insect pests and diseases is tied to optimal physical, chemical and mainly biological properties of soils. Soils with high organic matter and active soil biology generally exhibit good soil fertility as well as complex food webs and beneficial organisms that prevent infection. On the other hand, farming practices that cause nutrition imbalances can lower pest resistance" (Altieri and Nicholls 2003, 204). Synthetic fertilizer use also allowed farmers to plant crops closer together, creating fields that are more susceptible to pests. In addition to affecting pest populations through changes in soil and planting densities, the use of synthetic fertilizers also directly affects crops' physiological susceptibility to pests. Plants with boosted nitrogen levels from synthetic fertilization tend to benefit pest species. Indeed, a range of "crop pathogens, including fungi, bacteria, and viruses, also cause more severe damage when N inputs are high" (Matson et al. 1997, 507). All of these changes caused by synthetic fertilizer adoption help make synthetic pesticides a common tool for industrial agriculture.

To save their crops from disease and insect pests, often exacerbated by the local climate and synthetic fertilizer use, farmers in industrial agricultural systems turn to, synthetic pesticides as their go-to products for almost all control. This dependence is influenced by pests, their biology, and their ecological relations, but pesticide use is certainly not a natural response to the presence of pests and diseases; instead, much spraying occurs to meet often stringent market requirements around product cosmetic standards (Thrupp, Bergeron, and Waters 1995). Pesticide adoption can also become an economic necessity if farmers are to remain competitive with early adopters who can receive higher profits by beginning pesticide use before other farmers.

Pesticide adoption favors the interests of off-farm capital, which are those firms whose fortunes run through agriculture but do not formally participate in agricultural production, including input manufacturers and input retailers, creditors, landlords, and food industry businesses (Lockie 2006; Ward 1995).[4] Agrochemical manufacturers have aggressively promoted synthetic agrochemicals around the world (Bull 1982; van den Bosch 1980), creditors have commonly required farmers to use agrochemicals before providing agricultural credit for the season (Thrupp 1990b), and wholesalers and food retailers have increased their demands for beautiful and blemish-free produce (Freidberg 2009), which has prompted large increases in pesticide use for purely cosmetic purposes (Pimentel, Kirby, and Shroff 1993). Off-farm capital, then, strongly shapes the decision making of the farm family or farming business, both as input supplier and as enforcer (and oftentimes creator) of standards within the commodity chain.

Political economists have long studied how off-farm capital finds various entry points into agriculture, and pesticides are a prime example. Off-farm capital typically increases the circulation and accumulation of capital through agriculture by overcoming or modifying the natural processes underpinning agriculture, thereby making these heretofore farmer- or community-controlled processes part of the circuits of capital accumulation (Kloppenburg 1988; Mann and Dickinson 1978; Watts and Goodman 1997). Agrarian political economists call this industrial appropriation, whereby the inputs created and activities historically performed by farmers become appropriated by industrial and financial off-farm capital: "As elements of the rural production process become amenable to industrial reproduction, they are appropriated by industrial capitals and *reincorporated* in agriculture as inputs or produced means of production" (Goodman, Sorj, and Wilkinson 1987, 7). Kloppenburg's (1988) classic book shows how the fundamental starting point of agriculture—the seed—and its reproduction were made ripe for capital accumulation through two processes: the biological, through hybridization to increase yields and make subsequent generations much less productive, thereby forcing farmers to buy seeds anew each year, and the legal, which allowed for the patenting of plants, affording a state-enforced monopoly to exclusively benefit the creator of the patented seed.

As an example of capital overcoming natural barriers in agriculture, pesticide use shows how off-farm capital comes to profit from agricultural production processes. Costa Rican agriculture has been dramatically transformed through pesticide adoption as a form of appropriation, and these flows of off-farm capital through Costa Rican farming systems are complex and transnational. In the study site, these circuits of capital involve ninety-two different agrochemical firms and subsidiaries across multiple continents. These firms manufacture and formulate the 169 different kinds of pesticides available (table 1.3), which are made of 122 different active ingredients (shown by pesticide class in table A.1, appendix 2).[5]

Thus, as farmers adopt industrial inputs, their operations increasingly become part of these often long-distance circuits of off-farm capital accumulation. Banks, input manufacturers, and input retailers selling seeds,

Table 1.3 Country of manufacture and formulation for pesticides sold in Northern Cartago and the Ujarrás Valley

	Manufacturing location	Formulation location
Latin America	42	82
Brazil	5	5
Colombia	29	29
Costa Rica	4	32
Ecuador	—	1
Guatemala	—	9
Mexico	4	6
North America	35	35
United States	35	35
Europe	33	34
Austria	—	1
Belgium	6	8
Bulgaria	—	1
England	3	2
France	10	10
Germany	10	9
Italy	2	2
Switzerland	2	1
East & South Asia	10	10
China	2	3
India	1	1
Japan	6	5
Taiwan	1	1
Southwest Asia	7	8
Israel	7	8
Unspecified	42	—
Total	169	169

Source: Author's collection of Costa Rican pesticide label photos.

fertilizer, pesticide, and machinery have an increasing stake in farmers' production. Bell (2004) uses the example of Rob, a large-scale corn and soybean farmer in Iowa. Rob has a farming operation of six thousand acres, with two thousand owned and four thousand rented, with gross operating costs of $2 million per year; "a lot of people's fortunes funnel at least in part through Rob's" (Bell 2004, 93). Unlike farmers, these input manufacturing and retail sectors receive a return when they sell their product or service to the farmer, regardless of whether the farmer receives a higher price for his or her crop than the costs of production. Farmers keep all the risks of agricultural production, which are considerable, while agricultural input suppliers profit regardless of the outcome of production (this connects to theories of contract farming, which I highlight in chapter 2).

By theorizing pesticides geographically, we can see that industrial appropriation enables a spatial fix (Harvey 1999) for off-farm and on-farm capital by allowing for an expansion of market-integrated monocultural agriculture to areas where it was not previously possible, or at least not previously profitable. This spatial fix has interesting scalar dynamics and sets up a number of socioecological contradictions. In terms of the global flows of capital, as noted above, transnational pesticide companies have for decades attempted to increase sales in the developing world as markets became saturated in industrialized economies (Bull 1982; Weir 1987; Weir and Schapiro 1981). This search for external markets often coincided with development efforts aimed at promoting agricultural development through market expansion, market integration, and agricultural industrialization (de Janvry 1981). At the level of regions, pesticide use can control pests as the habitats of these pests' enemies—in the form of natural or field-edge ecosystems—give way to cultivated fields. Agrochemicals thus allow for new areas to come under production, including suboptimal areas for producing certain crops, and can keep these areas economically viable even as regional landscapes become less amenable to production. This, however, is a short-term strategy that only lasts until the conditions of production are undermined, as discussed below. Turning to the field level, pesticides allow larger-scale monocultures to be economically viable despite suffering from considerable pest outbreaks. In this way, capital finds a way to temporarily beat back pest pressure to allow for economical production relying on economies of scale that would not otherwise be possible because of ecological constraints.

In Northern Cartago, pesticides help farmers maintain a former cloud forest environment as a highly productive vegetable-production region. Farmers insist that pesticides are now essential for the production of many vegetable crops in the region to be economical, even though many farmers are certainly not happy about their dependency on costly and dangerous inputs whose price always seems to rise. This pesticide dependency helps drive the injustices of industrial appropriation: while becoming necessary for agricultural production (and therefore eventually losing their edge as a tool for increased profitability for individual farmers), all of the risks of production and pesticide use remain with the individual farmers, farmworkers, and rural residents, and the benefits largely accrue to off-farm capital. Pesticides also have a role in class differentiation in that they influence which farmers are able to survive economically in a competitive marketplace, as will be explored in chapter 2.

The appropriation of these agricultural and ecosystem processes by capital usually has profoundly negative environmental effects. James O'Connor (1993, 131), a leading theorist on capitalism and the environment, states his second contradiction of capitalism thus: "when individual capitals attempt to defend or restore profits by cutting costs, the unintended effect is to reduce the 'productivity' of the conditions of production hence to raise average costs. Cost may increase for the individual capitals in question, other capitals, or capital as a whole." Drawing on Marx, the conditions of production are things not specifically produced as commodities but are treated as if they were, including human labor power, the environment, and infrastructure; Polanyi (1957) calls these first two, together with money, "fictitious commodities." Marx (1990, 638), in using the case of English agriculture to understand the tendencies of capitalism, argued that "all progress in capitalist agriculture is a progress in the art, not only of robbing the worker, but of robbing the soil. . . . Capitalist production, therefore, only develops the techniques and degree of combination of the social process of production by simultaneously undermining the origin sources of all wealth—the soil and the worker." Marx's (1991, 216) discussion of capitalist agriculture ends with a broader conclusion: "The moral of the tale, which can also be extracted from other discussions of agriculture, is that the capitalist system runs counter to a rational agriculture, or that a rational agriculture is incompatible with the capitalist system (even if the latter promotes technical development in agriculture) and needs either small farmers working for themselves or the control of the associated producers."

O'Connor (1993) pushes further to argue that undermining the conditions of production eventually raises average costs for individual producers through the need to replenish nutrients in the soil through fertilizer use and for other producers through externalities from soil degradation, such as soil erosion, that impact others. Thus, updating Marx's line of thought to take into account the recent industrialization of agriculture and its inputs, O'Connor uses the example of pesticides to illustrate capitalism's second contradiction. "Chemical pesticides in agriculture at first lower costs but ultimately increase costs as pests become more chemical resistant and also as the chemicals kill the soil" (O'Connor 1993, 131; see also Wilson and Tisdell 2001). The production benefits of chemical pesticides are temporary because their common use leads to the pesticide treadmill, in which pests eventually develop resistance to pesticides, secondary pests are created as natural enemies are destroyed, and changes in soil qualities leave plants more vulnerable to pests. Farmers must spray more for the same or even a lesser level of control (van den Bosch 1980). And because these products are petroleum based, "farmers end up paying more and more for inputs that eventually degrade and destroy the environment into which they are being ploughed" (Kloppenburg and Burrows 2001, 108).

Studies in Northern Cartago have documented resistance to commonly used pesticides in many insect and disease pests. In addition to documenting pesticide resistance in a number of fungi and insect pests,[6] studies have found that bacteria have developed resistance to common antimicrobials used in human medicine—gentamicin, oxytetracycline, and streptomycin—that are also used as pesticides in Costa Rican vegetable production (Rodríguez et al. 2006). Chapter 3 shows how farmers have increased their spraying over the decades, likely in response to this resistance. The conditions of production are also undermined because pesticides compromise the viability of the soil (Thrupp 1991a), destroy pests' natural enemies and pollinators (Pimentel et al. 1992), and damage the health of farm families and workers (Antle, Cole, and Crissman 1998).

Appropriation tends to escalate as market conditions place farmers in strong competition against one another. The result, Willard Cochrane (1979) noted, is the "treadmill of production." Farmers must continually adopt the latest technologies, including pesticides, in order to remain competitive and continue farming, since failing to do so will make them less efficient relative to other farmers producing the same goods. Pesticide adoption, and eventually the pesticide treadmill, is an important part of the treadmill of production.

Yet while pesticides might have initially large benefits to individual farm units, these do not necessarily last. Once pesticides are adopted in order to remain competitive for economic survival, "it may be impossible to revert to the previous process, except at a high cost, even when the cost of production employing the new technique eventually rises above that of the old" (Wilson and Tisdell 2001, 457). In other words, even though pesticide use often becomes economically inefficient relative to previous or alternative practices—since it becomes less efficient over time by undermining the ecological conditions of production—it is very hard to change away from them and toward organic production systems, because the costs and risks of conversion are borne by individual producers who will be at a competitive disadvantage for many consecutive years if they convert (Tisdell 2005). Most farmers do not have a reserve of wealth ample enough to get through these years, the results are very uncertain, and they might also not have the knowledge to pull it off (Bell 2004). Thus, "farmers become locked into 'unsustainable' agricultural systems once pesticides are adopted . . . because of the initial heavy cost of switching to more sustainable systems and the need for all to act simultaneously in the switching process if economic losses are to be avoided" (Wilson and Tisdell 2001, 457–58).

Even when pesticides become a mandatory input as part of the treadmill of production, they do not guarantee farmers' livelihoods. Some farmers accumulate capital faster than others, and crop failure even with pesticide use can have a devastating effect on farmers, especially those who have been indebted by off-farm input purchases (van der Hoek et al. 1998, 501). In addition to being left with all of the production risks and being more heavily indebted due to input adoption, the farmers' share of the food dollar drops as agricultural industrialization through appropriation proceeds. In the United States, where good historical data are available, farmers received 41 percent of the value produced by economic activity in the agricultural sector in 1910 but only 9 percent by 1990. Conversely, input suppliers' shares went from 15 percent to 24 percent, and marketers' shares went from 44 percent to 67 percent (Lyson 2004, 58). We should expect similar outcomes where industrialization becomes commonplace in agriculture and concentration proceeds in the sectors of agricultural inputs and markets for agricultural commodities.

Overall, pesticides offer a Faustian bargain, locking farmers into a system of production that undermines the very conditions of production on which agriculture depends: the soil and its flora and fauna, surrounding ecosystems, and the bodies of workers and farmers. These developments make Marx's predictions about capitalism and agriculture quite prescient: "the way that the cultivation of particular crops depends on fluctuations in market prices and the constant changes in cultivation with these price fluctuations—the entire spirit of capitalist production, which is oriented towards the most immediate monetary profit—stands in contradiction to agriculture, which has to concern itself with the whole gamut of permanent conditions of life required by the chain of human generations" (Marx 1991, 754n27). The injustices for producers can be substantial, as pesticide adoption increases the share of value flowing to agrochemical input suppliers and thus reduces the share of value returning to the farmer while simultaneously compromising their health and that of farmworkers and farming families. In short, the expansion of a capitalist agriculture—of which pesticides are now an integral part—threatens lives due to toxic exposures and also threatens farmer livelihoods through the treadmill of production and the inevitable loss of less-competitive farm households and by undermining of the ecological conditions of production. The expansion of capitalist agriculture has all of these negative social and environmental impacts on specific locales while enriching off-farm capital, especially agrochemical input companies, creditors, and the purchasers of agricultural commodities.

Having laid out the basic processes at work in agrochemically dependent agriculture and its expansion, I now turn to the analyze their workings within the study site. Chapter 2 examines these dynamics geographically, while chapter 3 examines them historically.

2 Socioeconomic Differentiation and Geographies of Nature

Nelson, the farmer who first described the cloud belt to me (chapter 1), is the largest farmer of minisquash in my survey. He resides and owns agricultural land in Cipreses but refuses to plant his vegetable crops in the Cipreses area because of the large quantities of pesticides needed. Instead, he plants export minisquash and other minivegetables in the distant fields of other farmers with whom he works *a medias,* a relationship in which Nelson is an off-site manager and provider of most of the capital. Nelson farms in the much drier and less cloudy area west of Cuesta La Chinchilla (see figure 1.1) and notes that he is able to substantially reduce pesticide use relative to growing in Cipreses. Nelson's plantings further afield range from Alajuela Province (west of San José, to the west of the study site) to Santa Cruz de Turrialba (to the east of the study site). Nelson's land-use arrangements are what I call a *spatial strategy*—intentionally shifting production from the cloud belt to areas better suited to vegetable production. This is an important dimension of agriculture in the study site and one that intersects with processes of socioeconomic differentiation and the intensity of pesticide use.

The minisquashes—the backbone of the local minivegetable export sector—were originally introduced into and tested in the Cipreses area in the 1980s (Breslin 1996) as Costa Rica was incentivizing export-oriented production through externally imposed structural adjustment programs (Edelman 1999). Finding the conditions of production challenging, wealthier minisquash farmers from Cipreses eventually experimented with growing the vegetable in other locations and learned that Cipreses is a relatively poor location for minisquash because of the cloud belt. One of the two minivegetable exporting companies located in Cipreses now prohibits its farmers from growing minisquash in the area because of the heavy spraying required, which might translate into high levels of residues that foreign inspectors would detect (see chapter 4). Thus, like Nelson, minivegetable export farmers who can afford to do so shift production to sunnier and warmer areas to the south and west of Cipreses, since this leads to reduced plant disease pressure and allows them to spray less.

This chapter examines these social and ecological dynamics of agriculture by focusing on pesticide use and socioeconomic differences of farm households living in the cloud belt. The focus is on those households growing minisquash and potato. I take the understanding of the area's physical and pest geography developed in chapter 1 and relate it to household resources to examine the interrelationships between political economic and ecological processes in Northern Cartago, including the likely consequences for differential capital accumulation, social inequality, and environmental pollution. Socioeconomic differentiation in the area has occurred, and will continue to occur, through the complex interaction of climate, plant diseases, access to resources, and competition. Let's look at these interactions.

Socioeconomic Differentiation in Agriculture

Social and economic differentiation among groups of relatively homogenous agrarian landowners is a standard outcome of the development of capitalist agriculture (Goodman and Redclift 1981). Cochrane's (1979) landmark study, highlighting the treadmill of production in U.S. agriculture as discussed in chapter 1, shows that on-farm capital accumulation "is likely to be a socially and economically uneven process. Some farmers are more successful in accumulating capital so that the process becomes marked by increasing stratification or differentiation among farmers" (Buttel 1983, 111–12). This leads to what Cochrane calls cannibalism: "larger, more aggressive farmers outcompeting and purchasing the lands of their less successful neighbors" (Buttel 1983, 112).

The specific trajectory of differentiation in agricultural production has been debated since Lenin and Kautsky's exchange about the peasantry's fate with agrarian capitalism's expansion. Lenin argued for a traditional trajectory of concentration following Marxian thought in which capitalist farms would outcompete and eventually eliminate the peasantry as simple commodity producers (social units of production organized by households). This would create a rural bourgeoisie and a rural proletariat, which might still have a small plot of land but whose social relations would be more that of a worker (de Janvry 1981, 99). Kautsky disagreed based on empirical evidence and noted that the peasantry, despite being economically inefficient, continued to persist for a number of reasons (Banaji 1980). Kautsky's argument was prescient: while in most other economic sectors simple commodity producers have been outcompeted by capitalist firms producing the same commodities, in capitalist agriculture, family farm units compete directly and often very successfully with corporations. In some cases, family farms remain the only units of production in the sector. This makes what Kautsky called the *agrarian question*—the curious persistence of family farms with the spread of agrarian capitalism—continually relevant (de Janvry 1981; Marsden et al. 1996; Watts and Goodman 1997).

The ways that the agrifood industry has utilized contract farming and other noncapitalist labor arrangements to its advantage feature prominently in explaining the agrarian question. For example, contract farming—in which food industry firms contract with farmers to receive their agricultural products rather than going into production themselves—allows off-farm capital to pursue capital accumulation through agriculture while avoiding the many risks inherent in direct ownership and control of the means of agricultural production. With farmers tied into the larger food system through contracts, farming becomes agricultural piecework[1]—"a set of relationships in which labor (disguised as independent owner-operators) makes few production decisions" and "retains the fixed costs and risks of production while managing its own self-exploitation" (DeLind 1991, 41). The endpoint of this argument is that family farmers who enter into contract farming relationships are "propertied proletarians" with little to no say over production decision making (Lewontin 2000). Thus, while off-farm capital does not take direct possession of the means of production, it exerts enormous influence over agriculture through contracting, as it does also through the mechanisms of indebtedness and monopoly control of inputs. In this way family farms "become an integral part of capitalist agricultural development—instead of a barrier against it" (Davis 1980, 146). Overall, then, family ownership is quite "functional to external [off-farm] capital" (Marsden 1988, 319), which helps explain its persistence.

Another explanation of the persistence of family farming is the ability of farm families to self-exploit to remain competitive with capitalist farming firms (Banaji 1980; Chayanov 1966). Farm families can reduce consumption in hard times and persist, even as capitalist firms go out of business because they cannot remain profitable. Additionally, family labor power need not receive a wage in return, in contrast to wage laborers employed by corporate farms (Friedmann 1978), thereby providing another competitive edge for family farms.

Nature as a barrier to capital accumulation has also featured prominently in the agrarian question. Mann and Dickinson's (1978) influential thesis discounts self-exploitation and argues instead that capitalist firms do not dominate agricultural production directly because of a difference in socially necessary labor time and the production time that makes agriculture unattractive to capital's direct control. Socially necessary labor time refers to the amount of human work time required under the average conditions of production to produce a commodity. Production time refers to the time when an unfinished commodity, such as a plant or livestock, is "abandoned to the sway of natural processes" while not being within the labor process; "while labour generally initiates these processes, after this initial labour input the process proceeds on its own" (472).

Areas of production with a wide disconnect between production time and labor time—including most of agriculture—remain unattractive to capital on a large scale and thus remain in the hands of simple commodity producers. This is broad socioecological theory without a great deal of empirical backing. Mann and Dickinson (1978, 478) acknowledge this: "The theoretical approach which we have sketched out here must only be used in conjunction with a social and historical analysis. That is, the theory does not stand alone, but rather is modified by concrete historical processes; therefore, the production of each commodity must be examined in its historical, social and political setting." For example, an interesting question that arises is whether we see increased investments in agricultural production when returns in other sectors of the economy drop. If so, this would mean that nature as obstacle is not fixed but instead is relative to other opportunities in the economy. Along similar lines, Boyd, Prudham, and Schurman (2001, 560) provide a useful modification of the nature as obstacle thesis: "natural obstacles can also become opportunities for competitive advantage based on strategic and competitive mobilization around such obstacles, as well as on attempts to overcome them."

Nature as obstacle to and opportunity for capitalist development are but two ways in which nature intersects with production systems. Political ecology has paid more attention to the other ways that nature matters. For one, political ecology commonly focuses attention on important environmental outcomes—soil degradation, deforestation, pesticide contamination, etc.—of socioeconomic processes. But political ecologists have also argued that social processes, including socioeconomic differentiation, are *mediated through* environmental processes and change rather than these processes acting only as a barrier or an opportunity for capital. For example, Blaikie (1988, 144), drawing on Johnson, Olson, and Manandhar (1982) and discussing land degradation in Nepal, argues that "differences in resources between farmers may lead to differences in net damage [from mass wasting] suffered by the poor and the wealthy, so that 'the overall effect of "random" landslides and floods may result in increased disparities between rich and poor.'" More broadly, political ecology's well-articulated degradation and marginalization thesis (see the Introduction) links processes of socioeconomic differentiation and environmental degradation by invoking the concept of marginalization—and its social, economic, and environmental aspects—by arguing that those producers marginalized socioeconomically are pushed into marginal and fragile environments where land degradation may proceed rapidly, since these producers can ill afford investments for preventing it. This downward spiral means that socioeconomic marginalization and environmental degradation are dialectically produced.

Political ecology, then, has long argued that a differentiated nature has causal powers in socioeconomic differentiation processes. Moore (2008, 61) goes a bit further, calling for "situating ecological relations *internal* to the political economy of capitalism—not merely placing concepts of ecological transformation and governance *alongside* those of political economy, but reworking the fundamental categories of political economy from the standpoint of the historically existing dialectic of nature and society." Yet there exist few empirical studies on the dynamics of agricultural systems that pay particular attention to how these interactions—for example, between socioeconomic differentiation and ecological processes and relations—occur (but see Robbins 1998, 103).

This chapter takes up this charge and examines the ecological relations internal to agrarian capitalism in the study area. We start with the geography of pesticide use, which shows important differences to be explained.[2]

The Geography of Pesticide Use

Pesticide use in the area shows substantial geographic variation. Figure 2.1 shows the spatial distribution of pesticide intensity in export minisquash, imposing data from field-specific crop spraying schedules that were created by data gathered through the farmer survey (see Appendix 1) on the cloud belt from chapter 1. Farmers' commutes are shown with arrows, which represent "as the crow flies" rather than actual use of roads. Table 2.1 compares pesticide intensities of export minisquash grown inside and outside of the four-hour isohel that represents the cloud belt. Export minisquash is significantly more heavily sprayed when produced in the cloud belt, and fields outside the cloud belt in the lower elevations to the south and west are significantly less pesticide intensive. Three farmers also grow some or all of their export minisquash in areas outside of the study site (off of the map) to reduce pest and pathogen pressure.

Potatoes are also significantly more heavily sprayed when grown in the cloud belt. Figure 2.2 shows the geography of pesticide-use intensity for potatoes produced for the national market, using data from field-specific crop spraying schedules. While strong patterns are not immediately evident in the map, there is a decrease in average pesticide intensity as one increases elevation (moves away from the cloud belt and the high relative humidity that it represents). Table 2.1 compares pesticide intensities of potato production inside and outside of the four-hour isohel that represents the cloud belt. Potato production in areas to the north and west of the cloud belt is significantly less pesticide intensive than production within the cloud belt.

The geographic differences in potato pesticide intensities become still clearer when the considerable variability in pesticide use between farmers is controlled for in the analysis. There are twenty-two farmers in the

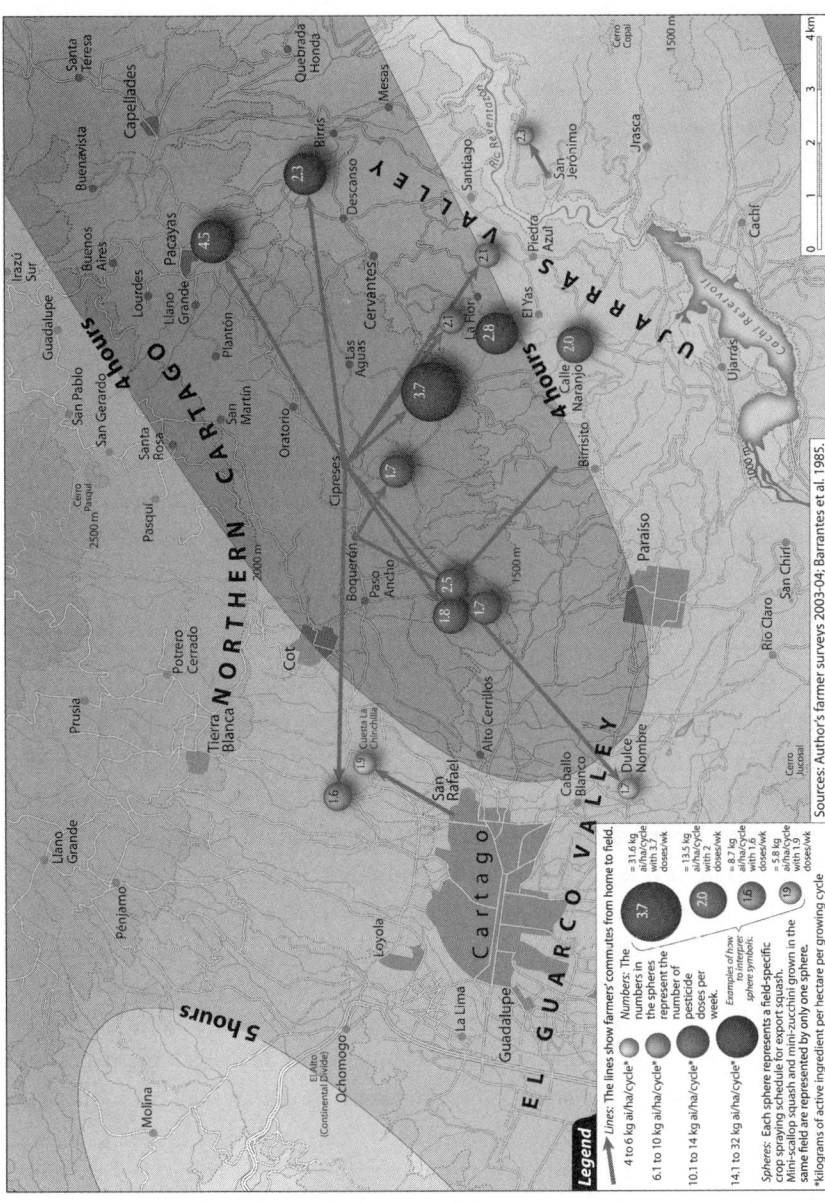

Figure 2.1 Export squash pesticide intensity and the cloud belt, Northern Cartago and the Ujarrás Valley.

Table 2.1   Minisquash and potato pesticide intensity in relation to the cloud belt, Northern Cartago and the Ujarrás Valley, 2003–2004

Location	Minisquash (kg ai/ha/cycle)			Potato (kg ai/ha/cycle)		
	Mean	St. Dev.	n[a]	Mean	St. Dev.	n[a]
Grown inside cloud belt[b]	15.9	7.3	16	63.8	31.3	55
Grown outside cloud belt[c]	6.9	2.6	11	48.9	26.9	42
Two-tailed t-test, p=	0.00			0.01		

[a] Data is from field-specific crop spraying schedules. Some farmers contribute more than one to the sample size if they grow both scallop squash and zucchini and/or farm in different locations.
[b] Annual average of less than 4 hours of sunlight/day.
[c] Annual average of more than 4 hours of sunlight/day.
Source: Author's farmer surveys, 2003–2004.

survey who grow their potatoes at different elevations. Nine farmers whose different potato fields are less than 380 meters apart in elevation (mean difference by farmer is 167 meters) reported no pesticide-use differences between their fields. However, for the thirteen farmers with two or more fields planted to potato that are more than 380 meters apart in elevation (mean elevation difference by farmer is 881 meters), twelve of the thirteen farmers' pesticide intensities drop substantially with increased elevation. For these farmers, producing at higher elevations north of the cloud belt results in a 29 percent reduction in pesticide intensity on average, as measured by kilograms (kg) of active ingredient (ai) per hectare (ha) per cycle, which is an absolute average decrease of 18 kg ai/ha/cycle. For pesticide doses per week there is a 42 percent reduction, from 4 doses per week in lower fields to 2.3 doses in higher fields. These reductions occur despite an increase in the length of the crop cycle by twenty-four days on average. That is, even though the crop cycle is longer, allowing more time for more pesticide use, most farmers still decrease their pesticide use with a shift of production to a higher elevation outside of the cloud belt.

As noted in chapter 1, the scientific literature on pests and agronomic conditions agrees with farmers' understandings that moving into the sunnier and less humid areas outside of the cloud belt can reduce disease

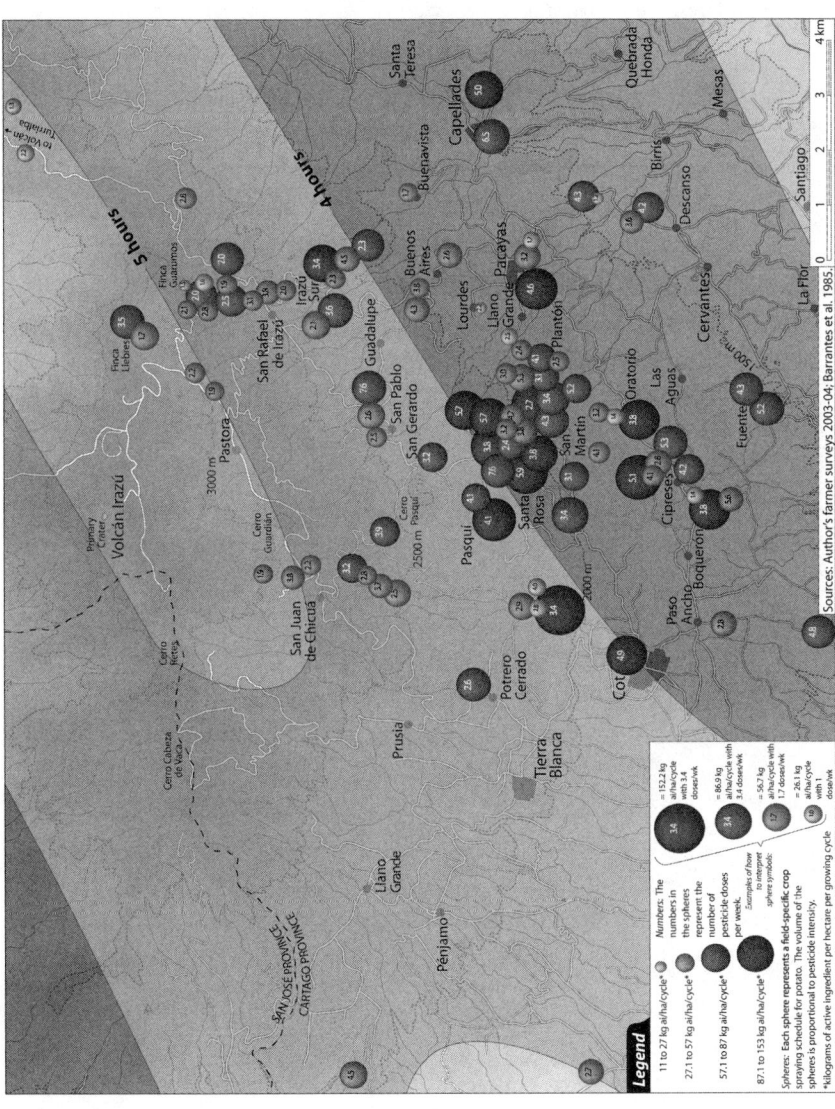

Figure 2.2 Potato pesticide intensity and the cloud belt, Northern Cartago.

pressure and pesticide use, especially fungicide use. The data above further corroborate these understandings.

To consider change over time, I asked whether farmers had increased or decreased their pesticide use over the last three years on specific crops. Figure 2.3 shows the reported change for farmers in the groups discussed above. All export minisquash farmers who live in the cloud belt and farm outside of it had reduced their pesticide use, while those growing in the cloud belt either maintained or increased their pesticide use. Decreasing pesticide use over time is a rare response in the study site. For potato

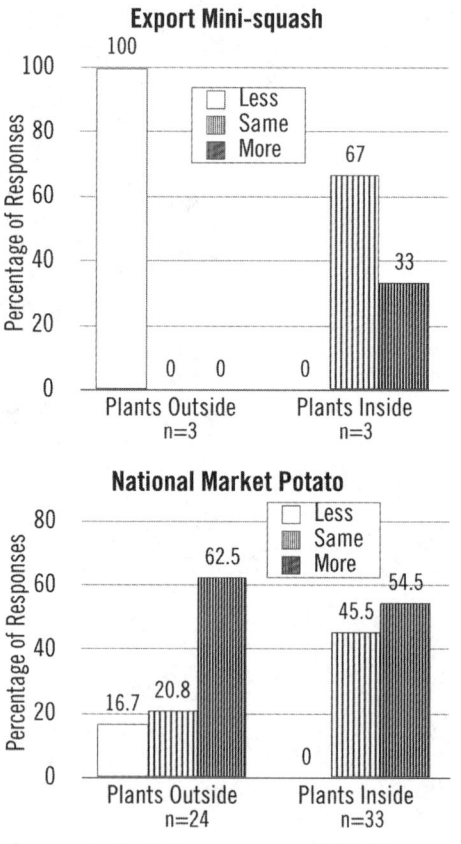

Source: Author's farmer survey 2003-04.

Figure 2.3   Change in pesticide use in the last three years, by planting location relative to the cloud belt.

farmers the results of pesticide-use trajectories are more mixed, but there are differences between the groups. Fifty-five percent of farmers who grow and live in the cloud belt have increased their pesticide use in the last three years, and none have decreased their pesticide use. In comparison, 63 percent of those living in the cloud belt but growing outside of it have increased pesticide use, but 17 percent have reduced their pesticide use. Pest pressure appears to be getting worse for potato farmers, but only those outside of the cloud belt have been able to reduce their pesticide use. With both crops, the group of farmers accessing land outside of the cloud belt are significantly more likely to have reduced pesticide use in the last three years.

Farmers' Resources and Environmental Advantage

Since growing outside of the cloud belt can significantly reduce pesticide use, why don't more farmers use this spatial strategy? The short answer is deeply political ecological: access to resources. Those minisquash farmers using a spatial strategy—farming outside of the cloud belt despite living within it—have significantly more wealth in their asset portfolios. The left-side columns of table 2.2 compare asset portfolios for two groups of export minisquash farmers who live in the cloud belt. These groups are (1) farmers who move their production out of the cloud belt (to the west and/or south) and (2) farmers whose production remains in the cloud belt. All measures of wealth are higher for the first group, and home, land, and tractor ownership are all significantly greater. Unfortunately, profit rates are not representative because only two of five farmers using a spatial strategy reported their total agricultural sales during the survey, although I suspect that the three withheld the information because their profit rates were very high. Thus, minisquash farmers who use a spatial strategy of moving outside of the cloud belt are generally more resource-rich than farmers who do not, and they also have significantly lower pesticide intensities in the production of their export minisquash.

For cloud belt households that do not move production and farm only within the cloud belt, many face investment poverty. That is, they do not have "the ability to make minimum investments in resource improvement

Table 2.2 Asset portfolios of minisquash and potato farmers with residence inside the cloud belt, Northern Cartago, 2003–2004

		Minisquash			Potato		
		Production outside Cloud Belt	Production in Cloud Belt		Production outside Cloud Belt	Production in Cloud Belt	
		n=5 farmers[a]	n=7 farmers	p[b]	n=24 farmers[a]	n=33 farmers	p[b]
Number of houses owned	mean	1.6	0.86	0.09	1.05	0.87	0.04
	st. dev.	(1.34)	(0.38)		(0.38)	(0.34)	
Total value of house(s) owned[c]	mean	$54,688	$22,857	0.43	$26,307	$21,781	0.20
	st. dev.	($30,352)	($24,170)		($18,116)	($27,152)	
Land owned (hectares)	mean	6.5	0.66	0.03	9.5	2.17	0.03
	st. dev.	(6.96)	(0.98)		(15.44)	(3.97)	
Current land planted (hectares)	mean	6.95	3.95	0.15	7.14	2.79	0.05
	st. dev.	(5.63)	(3.1)		(4.2)	(3.79)	
Pickup truck ownership	mean	100%	86%	0.21	100%	61%	0.00
	st. dev.	(0)	(0.38)		(0)	(0.5)	
Tractor ownership	mean	40%	0%	0.04	26%	3%	0.00
	st. dev.	(0.55)	(0)		(0.45)	(0.174)	
Formal credit received, last 12 months	mean	$9,688	$4,833	0.14	$6,808	$5,915	0.30
	st. dev.	($5,984)	($4,091)		($4,044)	($4,077)	
Agricultural expenses, 2002	mean	$32,083	$10,382	0.12	$22,163	$7,195	0.02
	st. dev.	($17,515)	($8,022)		($20,646)	($8,145)	
Agricultural profit, 2002	mean	$3,750	$1,768	0.35	$9,927	$1,110	0.04
	st. dev.	($12,347)	($4,091)		($12,151)	($5,525)	
Rate of profit[d]	mean	11.7%	17.0%		44.8%	15.4%	
Pesticide intensity (kg ai/ha/cycle)	st. dev.	7.39	15.5	0.00	46.14[e]	65.91	0.00
	mean	(2.08)	(8.37)		(19.03)	(31.14)	

[a] For minisquash, n=3 for farmers reporting expenses and n=2 for farmers reporting profit. Sample sizes are too small for strong conclusions to be drawn for these categories. For potato, the most resource-rich farmer producing outside of the cloud belt is excluded because expenses, profits, etc. are outliers compared to the other farmers in the group.
[b] Independent samples one-tailed t-test, equal variance assumed.
[c] All monetary values in 2003 U.S. dollars. Exchange rate used is 400 colones/U.S. dollar.
[d] Rate of profit is not representative for minisquash farmers growing outside of the cloud belt because only two of five reported total earnings for 2002, which is used to calculate profit.
[e] This average includes pesticide data for only those fields outside of the cloud belt, although many farmers in the first column have multiple fields inside and outside the cloud belt.
Source: Author's farmer surveys, 2003–2004.

to maintain or enhance the quantity and quality of the resource base, to forestall or reverse resource degradation" (Reardon and Vosti 1995, 1496). These households rent most of the land they farm, in stark contrast to the more resource-rich farmers who use a spatial strategy. For them, the average amount of land owned, 0.66 hectares, is one-tenth the amount owned by farmers who move their production out of the cloud belt, and they are less likely to own vehicles. These households do not have the assets required to capture the many advantages of moving production out of the cloud belt.

Using a spatial strategy in export minisquash production has worked best for four resource-rich farmers and one resource-poor farmer in the survey who are able to capture *environmental advantage*—a production advantage that accrues from the environmental characteristics of a certain locale. This chapter started with the case of one of the wealthiest farmers participating in the survey, Nelson, who refuses to plant within the cloud belt due to the need for heavy spraying. His son, who also lives in Cipreses but grows most of his export minisquash southeast of Cartago near Dulce Nombre, noted that environmental advantage more than makes up for the commute time and cost. Similarly, Nacho does not plant his export minivegetables in the Cipreses area, even though he owns land and lives there. Instead, he grows most of them on a rented farm west of Cartago near La Lima,[3] which he rents with another farmer (the other farmer was the only farmer who refused to be part of the survey). Nacho explained that his choice of the location was because of a better climate but also because of another landscape variable: other proximate farms.

> There, we are alone, we don't have traditional farmers [who grow potato, carrot, cabbage, and onion for the national market]. Traditional farmers, everything that moves in the field, they have to kill it. . . . When there are more farmers [in the area], it is more difficult to manage pests. . . . With *Liriomyza* [leaf miners] we don't have problems yet [in La Lima]. In contrast, if you plant green beans here [in Cipreses], you have to apply avermectin to control the larva. Because here it is exaggerated, [the population of] *Liriomyza*. There we don't have this problem because we are alone; there aren't farmers around. (Farmer survey, May 2003)

All of the farmers discussed above travel many kilometers to get away from the climatic problems of the cloud belt and have found land to rent or sharecrop. In contrast, some farmers already have land in better environments, so they do not explicitly use a geographic strategy but nonetheless benefit from environmental advantage compared to those growing near Cipreses. For example, Rubén lives in Cipreses because his wife is from there, but he is from the town of La Flor to the south. His family still owns land south of La Flor, and his only privately held land is just north of the town. These resources and an older truck allow him to grow in a climate that is slightly better than in Cipreses, which helps him lower his pesticide use as well as production costs compared to farmers nearer the center of the cloud belt.

The majority of export minisquash farmers do not employ a spatial strategy to decrease pesticide use because they lack sufficient capital and/or these social relationships. Many farmers remain close to Cipreses because it is not easy to farm in distant locations, since this requires more capital spent on fuel and vehicle upkeep and, as important, requires knowledge of and access to land for rent or relationships with farmers in distant communities with whom to sharecrop. Instead of a spatial strategy, many of these farmers rely on incorporating organic methods in order to reduce pesticide use, thereby enhancing the environmental advantage of the land they use. But they emphasize that due to the climate of the area, they must still rely on pesticides unless they can construct plastic-covered greenhouses, which are very capital-intensive.

The relationship between resources and using a spatial strategy is the same for potato farmers. The right side of table 2.2 compares asset portfolios of potato farmers who live in the cloud belt and (1) shift some or all of their production to the north and/or west outside of the cloud belt with (2) those who maintain all their production within the cloud belt. As with minisquash farmers, potato farmers who move production out of the cloud belt have more resource-rich asset portfolios. Home, land, pickup, and tractor ownership as well as number of hectares planted and agricultural expenses and profit are significantly higher for farmers who use a spatial strategy. As with the comparison of minisquash farmers, pesticide use is significantly lower for those potato farmers growing outside of the cloud belt.

Potato farmers employing a spatial strategy all reported that moving production higher on Volcán Irazú means a longer growing cycle, higher yields, and higher quality. The general rule is that the higher one can plant the better, since increasing elevation decreases relative humidity and lowers temperatures. Potato's susceptibility to late blight, *Phytophthora infestans* (discussed in chapter 1), is the main reason behind farmers moving outside of the cloud belt; late blight will not develop and spread as quickly in cooler and less humid areas. But planting at higher elevations also reduces virus pressure, since lower temperature reduces the prevalence of aphids that spread potato viruses (Hord and Rivera 1998; Meneses 1990),[4] and other insect pest problems are also less severe.[5] The only pest organism that seems to be favored by an increase in elevation is blackleg (*Erwinia carotovora*), since it "is a serious problem at altitudes over 2,500 m" (Jackson 1983, 104). In weighing the environmental influences on the major crop pests, producing at higher elevations confers substantial environmental advantage for potato production and allows for significantly reduced pesticide use and increased yields.

In relation to the spatial understanding of important environmental and pest gradients developed in chapter 1, the field-specific pesticide intensity data suggests a powerful rationale for farmers' spatial strategies for both minisquash and potato. In general, farmers who shift their production out of the cloud belt can capture environmental advantage, which includes (1) reductions in pesticide use and the concomitant production cost for all crops, (2) increases in produce quality and quantity, and (3) potential decreases in pesticide residue levels, which is more important as a driving factor for export minisquash farmers than for potato farmers, as we'll see in chapters 4 and 5.

Foremost in the logic of accessing other fields for environmental advantage is the reduction of production costs. Since all pesticide active ingredients are imported into Costa Rica, their price in Costa Rican colones usually rises faster than inflation, because Costa Rica's inflation rate is generally higher than the countries in which the active ingredients are manufactured. Reducing the number of pesticide applications saves on labor costs as well, since spraying is not mechanized in the area.

Some specific examples show how large these pesticide savings can be. Hugo grows potato in Plantón, San Rafael de Irazú, and near Volcán

Turrialba. He estimates that the fungicides and insecticides that he sprays (rather than those applied to the soil) cost him 400,000 colones (about $1,000) per hectare in the zone he defined as moderate late blight pressure from Plantón to San Rafael de Irazú. In contrast, it costs 200,000 colones (about $500) per hectare in his highest field (Farmer participatory mapping, January 2004). Jorge also noted that the difference was large: for sprayed insecticides and fungicides, he spends about 1 million colones ($2,500) per hectare in Cipreses and 715,000 colones ($1,788) per hectare in San Juan de Chicuá, which is much higher in elevation (Farmer participatory mapping, February 1, 2004). Thus, farmers can save more than $500 per hectare on direct pesticide costs alone by pursuing a spatial strategy that takes them out of the cloud belt.

Higher yields and quality is another environmental advantage of producing outside of the cloud belt. During *temporales*—powerful storms that blanket the area for days—the production of squash drops considerably. For example, Nelson, who now grows only outside of the cloud belt, noted that if he produces 1,000 kilos a week from his minisquash field in a relatively sunny week, during a *temporal* "there is not sufficient light; then it can be that I produce 100 kilos, that's it. It drops!" (Farmer survey, June 2003). For potato, farmers attribute improved tuber quality and quantity to the longer growing cycle at the higher elevations. According to Dagoberto, who grows in Cipreses and San Juan de Chicuá, at the higher elevations "the quality of the product is better" (Farmer interview, April 2003). Farmers noted that potato yields were typically larger at higher elevations. Dagoberto typically obtains a harvest of 17.1 metric tons per hectare (MT/ha) in the area around Cipreses and 20.7 MT/ha in San Juan de Chicuá (Farmer survey, September 2003). Another farmer, Fabián, noted that his yields are normally 23.7 MT/ha in Lourdes, at the edge of the cloud belt and 35.5 MT/ha in Guarumos, well above the cloud belt (Farmer survey, November 2003). Farmers noted that pesticide savings and higher yields and quality together more than offset the gas and vehicle maintenance costs of the commute. Thus, differentials between land rent and/or land price for land inside and outside the cloud belt do not fully reflect these differences in production.[6]

Finally, export minisquash farmers explain their spatial strategy as part of complying with exporters' concerns over pesticide residues (discussed in detail in chapters 4 and 5; see also the Introduction). Growing outside of the cloud belt translates into significantly lower pesticide use and likely lower residue levels as well.

In summary, resource-rich farmers benefit from landscape-scale geographic knowledge of Northern Cartago and surrounding areas and the resources to access areas outside of the cloud belt. This access allows them to exploit environmental advantage to increase produce yield and quality and to decrease pesticide use. These differences allow for larger profit margins for resource-rich farmers because they obtain greater benefits from production environments that are better suited to their crops, without having to pay much for the difference in productive power that these lands and their locations afford.

These dynamics contrast with common views of pesticide use and farm scale or farm wealth. In popular accounts of Latin America, it is the large-scale plantations that use pesticides intensively, while smallholders participate in a peasant economy that involves some production for market and for subsistence, which means low levels of pesticide use (Barry 1987). Throughout Northern Cartago and the Ujarrás Valley, smallholders use pesticides more intensively than do the infamously pesticide-intensive banana plantations in the Atlantic lowlands. The integration of these smallholders into national vegetable markets led to the capitalization of production, including, crucially, rising and now high land values that reflect the high-value crops that can be produced. This translates into heavy dependence on pesticides to protect high investments in the crop, including labor, seed, fertilizer, and payments to land (in the form of rent or mortgage debt). Not spraying to protect these considerable investments amounts to short-term financial destruction, given the intensity of pest pressure in the area. Only a handful of the area's farmers maintain traditional plantings of corn and beans because of the much higher earning potential of market vegetables.

Within this highly capitalized farming system, the relationship between the intensity of pesticide use and the scale of the farming operation, or farm household wealth, is nondeterministic, as has been reported with other cases (Feder, Just, and Zilberman 1985). Within the cloud belt, wealthier farmers access production environments better suited to the production of their crop, thereby reducing pesticide use. But there are also larger-scale farmers in the area who use relatively high amounts of pesticides, despite having access to relatively good growing environments. In a related analysis, I have shown that there are no strong direct relationships between measures of farm household wealth, farm scale, and pesticide intensity in the area, and the weak relationships that do exist vary by crop. For example, the larger the squash field the less it is sprayed, holding constant a number of other socioeconomic, political economic, and agroecological relationships (Galt 2008c).

These findings go against agroecological theory but illustrate well that agroecological processes do not operate in a social vacuum; we saw above that wealthier squash farmers, who generally have larger fields, are more able to access better growing environments. Additionally, pesticides are divisible inputs, relatively cheap per dose, as opposed to nondivisible and expensive inputs such as a tube well or tractor. Even small farmers can afford pesticides, and using them on high-value crops makes economic sense. We need to move past a dualistic conceptualization of the relationship between farm scale and pesticide use because relationships within farming systems are complex and contingent, with the possibility that higher farm household wealth can lead to pesticide-use reduction in regions with variable environmental geography.

Degradation and Marginalization

One primary goal of political ecology is "to make rigorous and explicit the causal connections between the logics and dynamics of capitalist growth and specific environmental outcomes" (Watts and Peet 2004, 13). The fine-grained political ecological analysis presented above combines environmental geography (including pests ecologies), the geography of pesticide-use intensities, and household asset portfolios to provide insights into people-environment interactions in agricultural environments shaped by agrarian capitalism.

The findings have implications for political ecology's degradation and marginalization thesis. Most examples of degradation and marginalization come from production systems that are not particularly capital-intensive (Blaikie 1988), with tropical rain forests receiving considerable attention (Collins 1986; Durham 1995; Hecht and Cockburn 1990;

Painter 1995). As Collins (1986, 8) argues in the context of the Ecuadorian and Peruvian Amazon, "whether farmers shift toward monocropping in order to benefit from credit and technical assistance available to certain crops, intensify production to repay credit obligations or expenses associated with claiming land, or fail to use appropriate management practices because insufficient revenues force them to work off-farm, the impacts are similar. Shortened fallows and monocropping on the one hand, or failure to use appropriate technologies and practices on the other, lead to soil erosion and deterioration, declining yields, increased economic pressures and almost inevitably, loss of land." Socially marginal and investment-poor farmers in this system essentially work to clear the land, and then degradation leads them to abandon the land and typically turn it over to larger-scale farmers involved in ranching.

This chapter provides a different kind of case study by focusing on a capital-intensive, semi-industrial vegetable-production system. In Northern Cartago and the Ujarrás Valley, the precise poverty-degradation dynamic is different from those commonly noted in political ecology. This is because of the power and logic of two intertwined but functionally distinct circuits of capital: one for off-farm capital and one for farm households as simple commodity producers (and potential accumulators of capital).

Paradoxically, the circuits of off-farm capital flow more powerfully and profitably through those farmers who experience degradation and marginalization more profoundly. Resource-poor farmers who remain within the cloud belt spend considerably more on pesticides, commonly more than $2,000 per hectare, but often receive relatively low returns themselves, as these expenses cut into profit margins. While the agrochemical industry benefits, investment-poor farmers in the cloud belt face a vicious cycle and a bleak future: high and increasing pesticide use and the resulting degradation in the forms of reduction of the soil's biota and nutrient cycling capacity, widespread environmental contamination, and exposure of themselves and their families to toxins, together with relatively low or even nonexistent profit margins. Degradation and marginalization for these farmers, then, leads to greater enrichment of off-farm capital through increased agrochemical sales. These same circuits of off-farm capital also flow through resource-rich farmers' production system but less powerfully. Richer farmers, then, are able to capture more value from their production systems, leading to considerable capital accumulation for many of them.

This articulation of local climate and pest geographies with the circuits of off-farm capital strongly shapes socioeconomic differentiation in the area. By using a spatial strategy to take advantage of environments more conducive to their crops' needs, resource-rich farmers reduce agrochemical use intensity and production costs and increase profit rates. Resource-rich farmers therefore accumulate capital at higher rates. Thus, farmers with different levels of resources in the mid-elevations of Northern Cartago have different abilities to compete in capitalist agriculture.

Disaggregating these two types of capital—off-farm and farm-based—helps shed some light on the debates over nature as a barrier to or opportunity for capital accumulation in agriculture. For off-farm capital, including financiers, input manufacturers, and input retailers, pests and their ecologies are opportunities for appropriation, allowing for the expansion of capital circulation and accumulation into new realms, as noted in chapter 1. The ecological and human consequences—contamination, pest resistance, poisonings, cancer, etc.—are of little to no concern to off-farm capital. Indeed, quite perversely, pest resistance (one process undermining of the conditions of production) leads to greater sales and profits for off-farm capital, since farmers have to spray more or buy newer and more expensive pesticides as pests develop resistance and as pesticides compromise surrounding ecologies that support natural enemies. The effects of these ecological relations are quite different for on-farm capital: pests and their ecologies act as barriers to short-term economic success, which are temporarily overcome through adopting the off-farm inputs created through industrial appropriation. Nature as pests, then, is *both* an opportunity for and a barrier to capital accumulation, depending on where one stands in relation to appropriation.

Yet even nature as barrier/opportunity does not fully capture the dynamics of the environment-agriculture interface. For one, as Moore (2008, 59) argues, "many on the left have too long regarded capitalism as something that acts upon nature rather than through it." Additionally, the discussion of nature as barrier/opportunity does not consider organizational heterogeneity among the social units of production and the nonsocial causes of socioeconomic differentiation. Wells (1996, 308–9) argues that agriculture's grounding in nature helps explain the considerable heterogeneity in the organization of agricultural commodity systems: "This unevenness has to do in part . . . with the ways that agriculture's necessary and varying bonds to land and climate and its dependence on the not-entirely-overcome natural rhythms of biological processes affect the course of class relations. That is, agricultural producers are unevenly able to relocate to escape local social struggles or to achieve optimal pricing for production factors." Similarly, the uneven ability to relocate production to access environmental advantage is another driver of socioeconomic differentiation, as shown above.

In this chapter I have shown that agrarian capitalism works *through* heterogeneous pest ecologies and geographies. The ecological specificities of production help make sense of how socioeconomic differentiation proceeds. This means that differential profit rates, a main driver of socioeconomic differentiation in capitalist agriculture, result not only from a classical conception of differential rents created by land that is heterogeneous in soil and endogenous attributes, as David Ricardo argued (Barnes 1984; Ricardo 1971), but also from land that varies in its climatic characteristics, since these strongly influence pest incidence and pesticide use. Indeed, as pesticides have helped open up regions of production not previously hospitable to certain crops (let alone large monocultures of them), environmental differences that affect pest geographies become all the more important in the dynamics of agrarian capitalism. In this case, the environmental conditions of the cloud belt, together with unequal access to land outside of the cloud belt with environmental advantage, will likely continue to contribute to socioeconomic differentiation in the area.

In addition to identifying these contemporary dynamics of agrarian capitalism in the region, the analysis begs the question of how the production system became so pesticide dependent in the first place. The next chapter examines this environmental history.

3 An Environmental History of Agricultural Industrialization

Ernesto is in his seventies and still walks to his farm and works his land every day because he loves doing it. During our discussion, we sit in his high-elevation field overlooking the southern flank of Volcán Irazú, purple cabbages for national market dotting the dark, fertile, friable volcanic soil around us. While we talk, clouds periodically blow through from the east, finally enveloping us in a thick fog that is so common in the region. Like only a handful of other farmers I interviewed, Ernesto is old enough to remember farming in Northern Cartago before agrochemicals. According to Ernesto and other older farmers, the land was so fertile that one could plant potatoes, not fertilize or spray them, and have a beautiful and bountiful harvest. It sounds like a carefree existence compared to their dependence upon heavy and frequent applications of synthetic pesticides and fertilizers today.

Ernesto and other older farmers tell me that late blight, the primary scourge of potato farmers discussed in chapters 1 and 2, came to the area in the mid-1940s when synthetic fertilizers were first introduced. Although I could not find the details of how and when late blight arrived in Costa Rican libraries, I suspect that late blight accidentally and unceremoniously arrived on new potato seed tubers brought into the area in the 1940s as farmers adopted new practices and varieties to increase output. Before the 1940s, Costa Rican farmers did not have to fight late blight, enjoying a honeymoon period of relatively easy production (see Dark and Gent 2001) before the powerful and devastating disease caught up to its host. This is similar for other insect and disease pests of the area: there are often delays between the introduction of a crop and its enemies.

Even though Ernesto disapproves of the area's heavy dependence on agrochemicals, his farming is utterly dependent upon them, and he sees no practical way of changing. Ernesto and most farmers in the area insist that they have to spray to produce vegetables, especially to be of high enough quality for the national market produce buyers, and spray frequently they do. Before synthetic pesticides, farmers of the area prayed to San Isidro, who was said to drive off insect pests, especially the damaging cutworm, which kills newly sprouted seedlings. Since then, massive landscape change from cloud forest to vegetable fields, the onslaught of introduced pests and an amenable climate for them, and market integration and quality requirements have all driven farmers into the Faustian bargain of pesticide adoption.

As the interview with Ernesto and other older farmers and my library research made clear, Northern Cartago and the Ujarrás Valley have not always been pesticide-dependent regions. How does a production region become so dependent upon agrochemicals, to the extent that pesticide use is considerably higher than on the same vegetables in industrialized nations and such that farmers see no viable alternatives to applying toxic chemicals, often near their homes? The historical question matters, because ahistorical narratives typically accompany pesticide-use studies. Generally, scholars have argued that in Central America, export production caused grave ecological disruption in the form of ever-increasing agrochemical inputs (Conroy, Murray, and Rosset 1996; Murray 1994), an argument that aligns well with political ecology's degradation and marginalization thesis in which degradation increases as farmers integrate into regional and international markets (Robbins 2012). I call this story into question, arguing for more focus on farmers' integration into local and national markets because this market expansion can also lead to very high levels of pesticide use.

This agroecological history focuses on four key topics: crops (especially potato as the first market vegetable crop), pests, agrochemicals, and vegetable markets. I draw on information collected through interviews with farmers as well as library research, including primary and secondary sources and many theses produced for Costa Rican universities. Of all the vegetables produced in the region, researchers and the state have focused most of their attention on potatoes, so this crop takes pride of place in my analysis. Extension and academic publications and even basic data on other national market vegetables before the 1970s are rare, but

I use available sources as much as possible to expand the story to other national market and export vegetables.

From Staple Crops to Vegetables for National Market

Quircot, Ujarrás, Orosi, Tobosi, and Cot (see figure 1.1) formed one of the densest centers of indigenous population upon Spanish settlement of what became Costa Rica (Bolaños et al. 1993). Indigenous peoples grew corn, beans, and chayote for thousands of years in the study site, passing down the seeds and techniques from generation to generation and clearing fields in the cloud forests and rain forests of the area. Planting of these indigenous crops continued through the intermarriage of the indigenous people and the Spanish colonists that created the mestizo population of the area. Today, only a handful of farmers continue to grow the traditional varieties of corn and beans, and their acreage is minute relative to vegetables for market, while chayote has become a major vegetable export crop in the region.

The Spanish conquest radically transformed Northern Cartago and the Ujarrás Valley, especially its agriculture. Cash crops became important as market integration brought households into the monetary economy. Sugarcane production and processing and beef cattle rearing became important economic activities. Coffee production, which transformed Central American economies and ecologies, began around Cartago in the early 1800s (Ceciliano Romero 1986) and in the Ujarrás Valley in 1825 (Bolaños et al. 1993, 230). These new activities—sugarcane, cattle, and coffee production—formed the early backbone of Costa Rica's colonial export economy, supplemented decades later by bananas from the lowlands. While cattle production eventually transformed into dairying in the region, sugarcane and coffee remain important today around the Ujarrás Valley.

The first major landscape transformation involving introduced vegetable crops occurred in the early 1900s. Market-oriented vegetable production—for the national market—began with the first commercial planting of potato in 1910 near San Rafael de Oreamuno, now a suburb at the northeastern edge of the city of Cartago (Ramírez Aguilar 1994, 419). Between 1915 and 1920, farmers introduced potatoes into higher elevations around Cot, Tierra Blanca, and Potrero Cerrado. The potato varieties planted at the time were known as *criolla, morada blanca,* and *morada negra.* Eventually farmers brought the potato farther up the slopes of Volcán Irazú until it reached San Juan de Chicuá, near the peak (Ramírez Aguilar 1994, 419). As it spread, farmers displaced their staple crops to grow potato for market. As noted above, older farmers emphasize that in the early decades, and in contrast to the current situation, farmers did not need to apply fertilizer or spray pesticides (see also Ramírez Aguilar 1994, 420).

By the 1930s, potato and corn dominated the annual cropping land-scape of Northern Cartago, with some bean production remaining for local consumption. Having displaced staple crop acreage, potato was the preferred crop of the region because of its economic return. The canton of Oreamuno, which spans most of Northern Cartago, acted as the potato "'granary' to the central markets" (La Tribuna 1934, 15, cited in Arrieta Chavarría 1984). Farmers planted two potato crops a year, generating per hectare profits higher than the famously profitable coffee areas of Alajuela and Heredia to the west. Other vegetables were also produced in the area in the 1930s, "but this business was not of great importance . . . because of the low prices quoted in Cartago and San José" (La Tribuna 1934, 15, cited in Arrieta Chavarría 1984).

Other forms of market integration occurred in the 1920s and 1930s. While potato has remained locally important since 1910, other crops came and went quickly. A black bean boom and bust occurred in the Ujarrás Valley between 1925 and 1935 (Bolaños et al. 1993, 231). There was also a brief integration into export markets with banana produc-tion in the Ujarrás Valley between 1930 and 1940. The local company Saborio y Ulloa first purchased bananas and transported them to the Caribbean on the railroad line running through El Yas but had to give up its business when demanded by the United Fruit Company, one of the powerful U.S.-based companies operating in Central America that wanted the region for itself. When banana diseases became a problem locally, United Fruit abandoned the area, effectively ending commercial banana production there and dashing early hopes pinned on potentially high returns from banana exports (Bolaños et al. 1993, 229–30).

Dairy farmers in the area also became integrated into the national market by the 1930s. By that time, dairies in the region were produc-ing six thousand bottles a day and selling them in San José. One of the largest latifundios, Finca Coliblanco (covering the area from Pacayas to Irazú Sur, now somewhat parceled out through land reform into the area known as Buenos Aires) had 2,000 hectares planted to introduced grasses for dairy production, 450 milk cows, and 25 milking stable attendants (La Tribuna 1934, 15, cited in Arrieta Chavarría 1984).

Vegetable crops and dairying were in the region to stay. These perishable goods require good market access, as implied by the term "truck farming." The paved highway from San José to Turrialba that still runs through Cartago, Paraíso, and Cervantes was completed in 1936 (Morrison 1955, 207). The paved road from Cartago to Capellades running just south of Cot and through Cipreses was completed before the 1950s. These roads allowed for the rapid growth of truck farming and dairying in Northern Cartago and the Ujarrás Valley by connecting small farming communities to the very close main local markets of Cartago and Paraíso and to the national marketplaces in San José. The fertile volcanic and alluvial soils, abundant rainfall, early introductions of vegetable crops, proximity to the main markets, and paved roads allowed the region to become the most important vegetable-producing area of Costa Rica, one that would soon become highly dependent on off-farm inputs and credit. Other areas of the country, more remote and with less fertile soils and less consistent rainfall, persisted as centers of subsistence corn and bean production and thus were less attractive to off-farm capital.

Maps of land cover in 1951 from Cartago to Turrialba provide a valuable source of information on the agricultural landscape of the lower elevations of the study site during this time of increased integration into the burgeoning national vegetable market (figures 3.1 and 3.2). The maps cover an elongated triangle of land between Cachí, the eastern edge of the city of Cartago, and Capellades (see figure 1.1 for the larger context). They show that the lower reaches of Northern Cartago and the Ujarrás Valley were a mosaic of mostly pasture and sugarcane on the large haciendas; small fields of corn, potato, and other commercial vegetables on minifundios; and forest-covered ravines. Corn and vegetable production existed on small plots around Cervantes and even more so around Las Aguas and Pacayas. Morrison notes that higher on the volcano, off the map, "pastures are more verdant, and large fields of potatoes account for most of the national production of that crop" (Morrison 1955, 210).

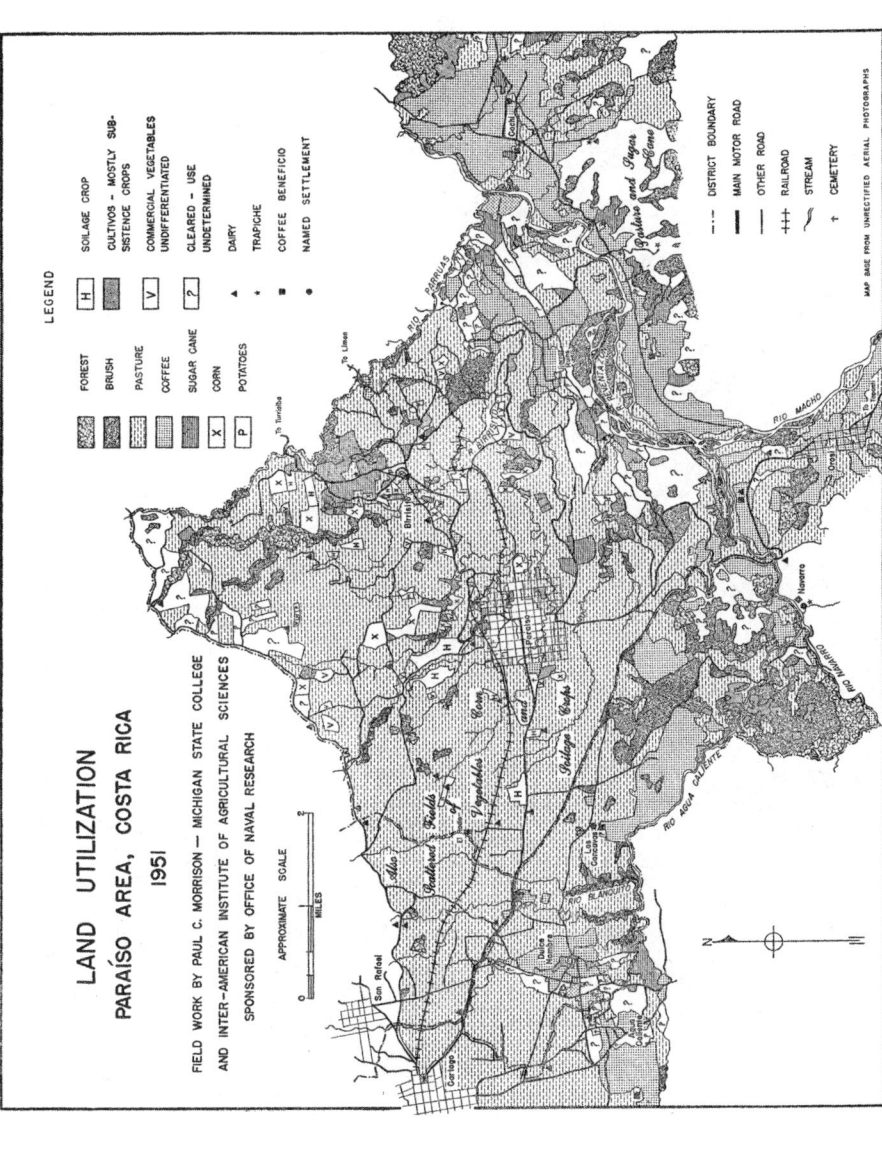

Figure 3.1 Land utilization, Paraíso area, 1951. (Source: Morrison 1955)

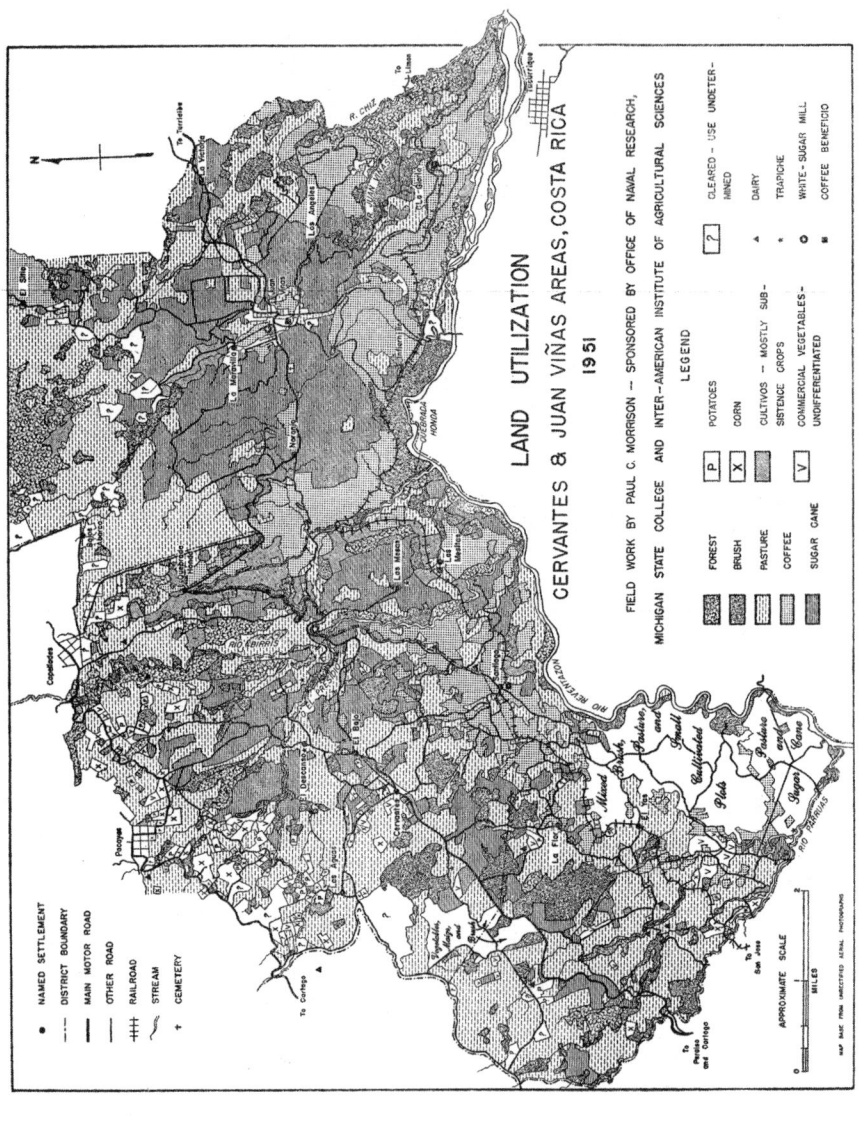

Figure 3.2 Land utilization, Cervantes and Juan Viñas areas, 1951. (Source: Morrison 1955)

Indeed, at the time, Northern Cartago produced an estimated 97 percent of the nation's potatoes and also had a large amount of pasture. In 1955, an estimated 66 percent of farmland in the Northern Cartago was in pasture (Lombardo n.d., 24, cited in Arrieta Chavarría 1984).

In the Ujarrás Valley, coffee, pasture, and sugarcane dominated, but farmers in the 1950s also devoted many small plots to commercial vegetables and some to corn and other staple crops. Unfortunately, chayote, the dominant crop currently in the valley, is not differentiated from other vegetables on the maps. Morrison's (1955, 213) only mention of chayote is in Quebrada Honda, the deep gorge west of the town of the same name, where he noted that it was "of particular consequence." His article also includes a photograph that shows a large field of trellised chayote planted on a steep hillside in Quebrada Honda. In reading the landscape in the photograph and looking at his map, it appears that he has symbolized the area of chayote as "cultivos—mostly subsistence crops." Since it is a large area, it seems likely that this was destined at least in part for the national market. This is supported by the fact that prior to chayote exports in the 1970s, farmers in the Ujarrás Valley filled sacks and loaded them on horses and wagons to take to the train stations of Santiago and El Yas to supply the national market (Bolaños et al. 1993).

By the 1960s, small farmers as simple commodity producers were heavily engaged in capitalist exchange relations in the area and in Costa Rica generally (Seligson 1980). According to the 1963 census, 80.8 percent of Costa Rican farms sold at least some of their produce (table 3.1). In Cartago Province the rate was higher, 82.6 percent, with most of the province's cantons having a higher proportion of farms selling their goods. Trejos's *Geografía Ilustrada de Costa Rica* describes the main crops of Cartago Province in the 1960s as "coffee, of very good grade from the valleys of Orosi, Tucurrique and Cachí; sugarcane, corn, beans, potato; magnificent vegetables and excellent pastures" (1966, 60). He then breaks down crops by canton, the administrative division below province (these production activities are also provided in table 3.2). The canton of Paraíso, which includes the towns of Paraíso, Santiago, Orosi, Cachí, Birrisito, El Yas, and Urasca (the southwestern side of the study site on figure 1.1), produced mostly coffee and sugarcane. Trejos makes no mention of chayote, which is one of the main crops today, even in

Jiménez Canton that includes Quebrada Honda, the location near which Morrison (1955) mentioned the importance of chayote in 1951. Instead, Trejos notes that cattle production, coffee, and sugarcane dominated. This is not to say that chayote was not produced there but that it was likely not a major crop. Costa Rican households have historically produced chayote in dooryard gardens, so the lack of concentrated areas of commercial production before export market expansion in the 1970s is not surprising.[1] However, detailed retail prices for chayote in San José date as far back as 1952 (Pineda Cabrales 1973), so it is likely that some farmers in the area were dedicated to commercial chayote production for national market since at least the 1950s (as noted above for Quebrada Honda).

Potato production in Northern Cartago remained strong in the 1960s; as Trejos (1966, 114) notes, potato provided farmers with "very good profits." The main crops of Alvarado Canton, which includes the towns of Pacayas, Cervantes, and Capellades, were potato and corn, with dairy farms and sugarcane as well (see table 3.2). Potatoes, vegetables, and pasture dominated Oreamuno Canton, which includes the towns of San Rafael de Cartago, Cot, Potrero Cerrado, Cipreses, and Santa Rosa. The exception was Cot, around which corn was still grown "on a grand scale" (68); it is interesting to note that Cot is referred to today as the most indigenous town of the area, so an affinity for corn even as national market integration proceeded is not surprising. Trejos also implies that a national market for vegetables (in addition to potato) had developed, with some regional exports: "Vegetables are grown on a large scale in the provinces of Cartago and San José, from where some quantity is exported to Panama" (114). The census data reflect these observations, with Oreamuno Canton having the highest percentage of dedicated vegetable farms in the area[2] (see table 3.2).

The 1970s witnessed a further increase in acreage devoted to national market vegetables in the study site. Castellanos Robayo (1972, 21) mentions that in 1972, corn production continued to decline in Northern Cartago and that carrot, beet, cauliflower, onion, cabbage, and chiverre[3] had become important crops. Farmers moved away from criollo corn production because of its high demand for soil nutrients and its growing cycle of eleven to thirteen months, which competed with more profitable

Table 3.1 Percentage of farms selling agricultural
goods, 1963

Province/Canton[a]	Farms selling goods	
	Percentage	n
San José	85.7%	15,262
Alajuela	83.6%	15,180
Cartago	82.6%	5,096
Central	84.9%	1,098
Paraíso	90.2%	562
La Unión	84.9%	238
Jiménez	82.9%	293
Turrialba	83.5%	1,814
Alvarado	77.5%	325
Oreamuno	83.4%	314
El Guarco	65.5%	452
Heredia	89.2%	3,088
Guanacaste	72.9%	10,773
Puntarenas	74.6%	9,941
Limón	80.2%	5,281
Costa Rica	80.8%	64,621

[a] Indented names below Cartago are the cantons of the
province.
Source: Dirección General de Estadística y Censos (1965,
244–45, table 208).

crops for space. Maps in his and other's work (Fuentes Madríz 1972)
show that the area from Cipreses to La Chinchilla was about equally
dominated by vegetable production and pastures for dairy cattle, a situation similar to today.[4] Pineda Cabrales (1973, 28) estimates that in the
Ujarrás Valley, in the early 1970s, chayote covered 171.5 hectares (245
manzanas) planted largely by small farmers, likely a substantial increase
from previous decades, since chayote acreage was not mentioned in
sources from the 1950s and 1960s.

By the 1980s, the conversion of annual cropping systems from staple
crops to national market vegetables was almost complete. While corn
dominated the area for millennia and was grown on 5,188 hectares in
Cartago Province in 1950 (Dirección General de Estadística y Censos

Table 3.2 Percentage of farms dedicated primarily to certain agricultural goods, 1963

Province/Canton[a]	Potato	Horti-culture	Coffee	Banana	Cattle/Dairy	Sugar-cane	Cacao	Rice	Beans	Corn	Other	n
San José	0%	0%	46%	0%	9%	2%	0%	4%	5%	4%	30%	15,262
Alajuela	0%	0%	33%	0%	18%	8%	2%	6%	5%	3%	25%	15,180
Cartago	4%	1%	45%	0%	17%	8%	0%	0%	1%	5%	19%	5,096
Central	9%	1%	50%	0%	11%	1%	0%	0%	1%	5%	22%	1,098
Paraíso	0%	1%	64%	0%	9%	11%	0%	0%	1%	0%	15%	562
La Unión	0%	0%	82%	0%	15%	0%	0%	0%	0%	0%	8%	238
Jiménez	0%	0%	31%	2%	8%	16%	0%	0%	0%	0%	41%	293
Turrialba	0%	0%	53%	0%	18%	13%	0%	0%	0%	6%	9%	1,814
Alvarado	15%	0%	6%	0%	24%	17%	0%	0%	1%	22%	16%	325
Oreamuno	23%	4%	2%	0%	27%	0%	0%	0%	2%	7%	35%	314
El Guarco	0%	0%	25%	0%	36%	0%	0%	0%	0%	1%	38%	452
Heredia	0%	0%	70%	2%	10%	1%	1%	1%	0%	1%	13%	3,088
Guanacaste	0%	0%	2%	0%	27%	0%	0%	11%	7%	9%	43%	10,773
Puntarenas	0%	0%	4%	1%	15%	0%	0%	16%	3%	5%	55%	9,941
Limón	0%	0%	1%	2%	5%	0%	43%	1%	0%	15%	32%	5,281
Costa Rica	0.4%	0.1%	26%	1%	15%	3%	4%	7%	4%	5%	33%	64,621

[a] Indented names below Cartago are the cantons of the province.

Source: Dirección General de Estadística y Censos (1965, 272–73, table 221).

1953, xvii), the area devoted to corn in the 1984 agricultural census had dropped to 1,207 hectares (Dirección General de Estadística y Censos 1986, 78, 80). No census data is available since 1984, but it is likely that the area used for growing corn is considerably lower today.

The Intensification of Agrochemical Use

The first annual crop transformation—from staple corn and bean pro-duction to national market vegetables—created the conditions for heavy pesticide use in the area. Farmers rarely use pesticides in milpa systems in other highland areas of Central America (Horst 1989) for two main rea-sons. First, criollo corn varieties are very resistant to insect pests and plant pathogens, according to the farmers growing them. Second, subsistence crops are grown mostly for use value (direct consumption) and generally have a low exchange value (the price they fetch on the market). Produc-ing a crop for use value generally means that farmers will not invest many expensive inputs into the system because there is no monetary return, al-though some farmers do use some pesticides to enhance subsistence har-vests if the crops produced by these inputs remain cheaper than buying the food or if market access to food is limited. In contrast, farmers use pesticides intensively when they expect a high economic return on a crop (Fernandez-Cornejo, Jans, and Smith 1998; Galt 2008b). In Costa Rica, the harvest from an average hectare of corn can be sold for $239, while the harvest from an average hectare of potato is worth $7,747 (Galt 2008b, 1387). This makes it uneconomical to invest in much agrochemi-cal protection of milpas. Corn farmers in highland Guatemala also make this observation: there are insects that eat corn, but they are not consid-ered pests because they do not cause significant economic loss (Morales and Perfecto 2000). In contrast, high-value crops such as potatoes take considerable investment, with farmers spending as much as $2,500 per hectare on pesticides (see chapter 1), a sum that would never be spent on pesticides for corn because of its low exchange value. Fertilizer inputs can increase yields and returns, and pesticides serve to protect the high-value crop that sits in the field and is subject to numerous pests. Like in Costa Rica, pesticide use per hectare in vegetables in Mexico is ten times higher than in corn production (Albert 2005).

The rise in simple commodity production—producing crops specifically for sale by farm households—can lead to regional crop lock-in. High returns on crops are capitalized into land values, driving land values higher (Guthman 2004). Returns to land are a necessary part of capitalist agriculture in the form of paying rent or servicing a mortgage. Land rental rates and land prices increase where high-value crops are produced, and farmers must then shift to these high-value crops to make rent or mortgage payments, unless farmers already own the land outright and property taxes are low (which is an important factor shaping the diversity in what is grown in a region). Thus, when farming regions are incorporated into a market allowing for high returns on crops, most farmers must adopt the high-value crops and will likely turn to pesticides to protect their investments in these crops. Farmers being in competition with one another, and thus facing the treadmill of production, also spurs pesticide adoption within this context (see chapter 1).

At the start of potato production in the area in 1910, it is likely that no pesticides were used in milpas. Inorganic pesticides—the only kind available at the time—were used exclusively on higher-value horticultural crops such as potato and orchard crops in temperate countries, where they were manufactured (Whorton 1974). Today the milpa system in the area, planted by only 3 farmers of 148 in my survey, has a pesticide intensity of 1.4 kilograms (kg) of active ingredient (ai) per hectare (ha) per year. Farmers plant it, spray a couple of doses of insecticide for cutworm (*Agrotis* spp.), and then leave it alone until harvest about a year later. This pesticide intensity is less than one-tenth the national average of 15.9 kg ai/ha/year, as calculated by Chaverri (1999). It is one hundred times less pesticide intensive than the annual potato-carrot-cabbage rotation, a national market annual rotation common in the area today. This provides an indication of the vast increase in pesticide use that accompanied expansion of national market vegetable production.

Potato production in Northern Cartago was followed a few decades later by the adoption of agrochemicals. La Tribuna (1934, 15, cited in Arrieta Chavarría 1984) makes no mention of pesticide use in the early 1930s but does provide some detailed information on harvest sizes and mentions fertilizer use, although this is likely organic fertilizers such as manure. Chemical fertilizers and pesticides were introduced into the area in the late 1940s by Rafael Cruz Mena (Ramírez Aguilar 1994), as was also noted by the oldest farmers in my survey. Farmers began synthetic fertilizer use as soil fertility declined, likely after years or decades of mining the soil by harvesting crops from fields without sufficiently replenishing soil organic matter and nutrient stocks.

Production for market also led to farmers continually seeking higher-yielding and pest-resistant varieties. Since the early 1900s, Costa Rican farmers have sporadically imported potato varieties without proper phytosanitary controls[5] (Sáenz Maroto 1955), although these controls increased greatly in recent decades. Late blight, the severe fungal enemy of potato in Northern Cartago, was introduced to the area a few years after the introduction of chemical fertilizers. Tuber moths (*Phthorimaea operculella*) were introduced early on as well. Another type of tuber moth, *Tecia* (*Scrobipalpopsis*) *solanivora*, now considered the most problematic Central American potato pest (Bosa et al. 2005), was introduced into the area from a shipment of potatoes from Guatemala in 1970. The pest caused a loss of 20–40 percent of the Costa Rican potato harvest in 1972 (Povolny 1973). Similar patterns of pest introductions likely occurred with other introduced vegetables, making farmers even more dependent on spraying to maintain their yields, although data on pest and pathogen introductions is generally lacking.

With the political, economic, and ecological conditions in place, chemical pest control, rare in most places in Costa Rica in the late 1940s, soon came to dominate agricultural production in the area (Arias Rodríguez 1998). By the 1950s, potato farmers relied on frequent applications of inorganic pesticides. A potato production manual (Sáenz Maroto 1955) provides detailed pesticide recommendations on potato and notes their common use. The manual describes farmers using mercury chloride, an inorganic compound, to disinfect seed potatoes. The manual recommends Bordeaux mix, a mix of copper sulfate and lime, as "indispensable to obtain good harvests" because of its action as a fungicide, an insecticide, and a foliar nutrient (Sáenz Maroto 1955, 5). This was to be applied at a rate of forty to seventy-five pounds per acre when the potato plants emerge, and then "3 or 4 sprays will be given every 7 to 8 days; the greatest lapse of time for spraying will be every 10 or 14 days" (Sáenz Maroto 1955). If the insecticidal action is not enough, the author recommended adding two pounds of calcium arsenate per fifty gallons of spray mix. He also notes that the synthetic fungicide maneb has helped greatly to control late blight in potato "and has been substituted in many areas for the use of Bordeaux mix" (Sáenz Maroto 1955, 25).

By the 1960s, synthetic organochlorine pesticides were in common use, and the search was on for late blight–resistant potato varieties. Molina Umaña's (1961) thesis evaluates new potato varieties' resistance to late blight in order to reduce fungicide applications. The very purpose of the research points to the pesticide-intensive nature of potato production in Northern Cartago at the time: late blight is "of great importance to Costa Rica now that the potato zone is rather extensive and its economy suffers disruptions that creates social problems" (1). It appears as though the inorganic compounds lost ground in the 1950s, since Molina Umaña sprayed his test plots with only the newer synthetic pesticides, which he notes were the pesticides commonly used in the area.[6] These included heptachlor as a soil insecticide at planting, maneb or mancozeb and captan for fungicides to control late blight, and DDT and methyl parathion as foliar insecticides to control tuber moth. He notes that his two experimental spraying schedules of every two weeks and every three weeks on late blight–resistant varieties in San Juan de Chicuá and Llano Grande were much less frequent than the at least weekly fungicide applications that farmers used in the area (13, 31).

In addition to this evidence that farmers sprayed insecticides and fungicides very frequently by the 1960s, their use also appears to have been widespread among the farming population. The 1963 agricultural census shows that in the three major potato-producing cantons of Cartago—Central, Alvarado, and Oreamuno—the use of fungicides is very common, with 83.2 percent, 76.3 percent, and 81.9 percent of farmers using them, respectively (table 3.3). This fungicide use is considerably higher than the average in Cartago Province (65.4 percent), almost double the rate in all of Costa Rica (44.1 percent), and more than triple the rate of the other cantons in Cartago that had little potato production (26.1 percent). The percentage of farms using insecticides in these potato-focused cantons, 43.7 percent, is also higher than the national average (37.6 percent) and more than double the rate of the non–potato-producing areas in Cartago (22.5 percent). These differences show the much more widespread use of

Table 3.3   Percentage of farms using pesticides, 1963

Region	Insecticides	Fungicides	Herbicides	n
Costa Rica	37.6%	44.1%	45.7%	4,431
Alajuela	37.6%	53.9%	34.6%	1,589
Cartago	37.6%	65.4%	42.9%	497
potato-producing cantons[a]	43.7%	81.1%	30.7%	355
other cantons[b]	22.5%	26.1%	73.2%	142
Guanacaste	31.2%	12.3%	85.1%	626
Heredia	42.8%	48.3%	53.9%	180
San José	49.9%	50.8%	17.4%	840
Limón	68.3%	23.7%	25.9%	139
Puntarenas	16.6%	26.6%	80.5%	560

[a] This is the average for Alvarado, Central, and Oreamuno.
[b] This is the average for El Guarco, La Unión, Jiménez, Paraíso, and Turrialba.
Source: Dirección General de Estadística y Censos (1965, 169–70, table 126).

insecticides and fungicides in the study site compared to the rest of the country, a response to the high value of their crop and the serious insect and disease pests facing potato farmers.

The intensive and widespread use of pesticides on potatoes has continued from the 1960s to today, and intensity of use has increased over time. Cost of production studies from 1971, done by Banco Crédito de Cartago, which offered agricultural credit throughout the country at the time, note that potato farmers in the lower areas spray twenty-four times per crop cycle in the dry season and thirty-two times in the rainy season, while higher-elevation potato fields were sprayed sixteen times per cycle (table 3.4). A 1975 publication on plant diseases in Costa Rica recommends fungicide applications of zineb or mancozeb every five to ten days on potato (González 1975, 4), which would be ten to twenty sprays per crop cycle. The use of synthetic fertilizer was also reported to be very high in Costa Rican potato production in the 1970s and early 1980s (Jackson, Cartín, and Aguilar 1981). Today pesticide applications are much more frequent, with an average of fifty-two to fifty-nine doses per cycle, compared to sixteen to thirty-two doses per cycle in the early 1970s (see table 3.4).

Table 3.4 Pesticide use on various vegetables, Northern Cartago and the Ujarrás Valley, 1971 and 2003

	Pesticide costs	Pesticide doses per cycle	
	1971	1971	2003
Beet	¢134	18	37
Cabbage	¢60	6	36
Carrot	¢217	19	50
Cauliflower	¢87	9	45
Corn & bean (*milpa*)	¢40	4	8
Green bean	¢30	2	21
Pea	¢68	8	46
Potato, Northern Cartago	¢601	24 to 32	59
Potato, San Juan de Chicuá	¢216	16	52
Tomato	¢768	25	71
Zucchini	¢76	12	29

Sources: Calderón Coto (1972); author's farmer survey, 2003–2004.

Information on other vegetable crops is much scarcer. Accounts of varieties planted, pest introductions, and pesticide use before the 1970s are unavailable for crops such as cabbage, carrot, onion, and most other vegetables. For example, for the brassicas in Costa Rica, "concrete information does not exist about the possible introduction date" (Saborío Mora 1994, 401). Some information on pesticide use is available from cost of production studies. In 1971, pesticide costs for potato and tomato in Northern Cartago were very high, while for most other vegetables it was considerably less (see table 3.4). Farmers were using pesticides on these other vegetable crops for national market but not nearly to the same extent as on potato and tomato, which are both severely affected by late blight.

Other sources show that farmers relied on agrochemicals to produce vegetables for the national market in the 1970s and 1980s. The earliest source discussing agronomic practices for chayote is Pineda Cabrales's (1973) study. He notes that households producing for subsistence used rotenone, a botanical insecticide, in their dooryard gardens. Commercial farmers controlled the major insect pests, especially mites and worms that bore into the fruit (*Diaphania hyalinata* L. and *Diaphania nitidalis*

Stoll), with the organophosphates ethyl parathion, malathion, tetradifon, methidathion, and diazinon. For nematode control, commercial farmers used ethioprophos, also an organophosphate. Pineda Cabrales (1973) also mentions the importance of four plant pathogens: powdery mildew, downy mildew, anthracnose, and mosaic virus. While Pineda Cabrales (1973) notes that chayote has a certain tolerance to these diseases, farmers at the time used Trimanzone (a combination of the fungicides ferbam, maneb, and ziram), maneb, and copper compounds to combat intense attacks of fungal diseases. Another source shows frequent pesticide use in celery in the Ujarrás Valley in 1980, at the rate of once per week in summer and twice per week in winter (Calderón Mata 1980, 24).

Farmers' experience with intense insect pest and fungal disease pressures on potato and tomato—especially after the arrival of late blight—likely led them to try out agrochemicals on their other vegetable crops produced for market, such as carrot, cabbage, and squash. More generally, though, a number of processes were working together for pesticide use to grow more prevalent and more frequent: crops' pests and pathogens were introduced; the production areas of these crops increased and land values rose; more farmers started relying on chemical pest control, making it economically necessary for other farmers to do so; impoverished soils and chemical fertilizers made crops more susceptible to pests; and the other ecological conditions of production supporting biological forms of control were undermined.

The Role of the State and Agrochemical Companies

Above I highlighted how potato production was pesticide intensive by the 1950s and discussed many of the processes leading to this outcome. Farmers adopted fertilizers to increase yields and profits and used pesticides to protect their investment in high-value crops, comply with market standards, and increase production; simultaneously, pesticide use was a response to changing agroecological conditions, especially introduced pests, the expansion of monocultures into inhospitable areas, and an undermining of ecosystem health. All of these agricultural transformations would not have happened without production for market, since farmers spray according to expected returns.

But farmers' economic logic in response to high crop values, high land values, and the agency of nonhuman actors are not the full story. Political-cal ecology suggests that powerful social actors, especially agrochemical firms, also shape land users' agrochemical use (Bull 1982; Robbins 2007; Robbins and Sharp 2003; van den Bosch 1980). Since the 1940s, states, development agencies, private foundations (at first most prominently the Rockefeller Foundation and now the Gates Foundation), international crop improvement centers, and agrochemical companies around the world acted in concert, advancing a modernization paradigm—for which the Green Revolution is used as shorthand—that embraced chemically intensive agriculture based on high-response varieties, as noted in the Introduction (Dinham 1991; Perkins 1990; Scott 1998; Shiva 1991; Wright 1990).

In Costa Rica, the state's main target of agricultural modernization until the 1980s was the nation's staple crops, especially corn and black beans as well as rice and potato. Prior to structural adjustment in the 1980s, the state pursued policies aimed at promoting crops of national importance for self-sufficiency and the persistence of the peasantry (Edelman 1999), similar to trends toward national food regimes at the time (Friedmann and McMichael 1989). The state created the Consejo Nacional de Producción (CNP) after the 1948 civil war with the mission of promoting basic foods production, including corn, beans, rice, potato, sorghum (as a livestock feed), beef, fish, and milk. While corn, beans, and rice received the most attention, from the 1940s to the 1980s the state attempted to increase the country's potato production, since potato was the third most important source of food in the country after rice and beans (Arrieta Chavarría 1984; Programa Regional Cooperativo de Papa [PRECODEPA] 1982, 7) and the eighth most important crop by value (Banco Central de Costa Rica 1982, cited in Segura Coto 1983).[7] The 1961 Phytosanitary Law (Ley de Sanidad Vegetal) No. 2852 describes the state's mission as related to potatoes: "the laws and norms that the State has issued in favor of the potato help the farmer to technify the crop so that it is much more profitable and has a superior technological level" (Programa Regional Cooperativo de Papa [PRECODEPA] 1982, 7–8).

This state-led technification of potato involved the continued and increased use of agrochemical inputs—already heavily used by the

1960s—and a very important program of certified, improved seed production that began in 1982 and continues today (Raub 2008). Rather than the introduction of a complete technological package, as was the case in much of the Green Revolution, the technification of Costa Rican potatoes was a gradual one, driven in part by farmers' predilection to experiment with new technologies in the face of competition and chemical companies' interests in increasing sales.

State-led potato technification focused on rectifying one of the largest problems that private capital would not tackle: the lack of improved and certified disease-free seeds with late blight resistance. Since potatoes are planted from seed potatoes (small tubers, not from true seeds), farmers at low elevations who use seed potatoes from their past crop often pass on the fungal, bacterial, and viral pathogens prevalent in these environments to the next crop, resulting in declining yields over time due to starting with diseased seed tubers. The Programa Regional Cooperativo de Papa (PRECODEPA) report lamented the "low" yields of 8.5 metric tons per hectare in 1976 resulting from a lack of improved seed—that is, high-response varieties certified to be free of disease—and emphasized the need for further technification of potato production with the seed program and increased extension effort (Programa Regional Cooperativo de Papa [PRECODEPA] 1982, 3).

Potato technification, intensifying in the early 1980s, worked well in terms of overall production and yields per hectare (figure 3.3). While stagnant from the early 1960s to the early 1980s, potato yields increased greatly starting in 1982. From Molina Umaña's (1961) work we can assume that pesticide applications were at least weekly since the 1960s. This may have increased in the 1980s with greater extension effort as suggested by the 1982 report, but it is difficult to tell since the yield increase may be largely the result of improved production from certified seeds. Much more likely, the use of fungicides and insecticides allowed farmers to continue to protect their crop, while the improved seeds allowed yields to take off as long as they were accompanied by synthetic fertilizer that farmers were already accustomed to using.

While state programs commonly have a written record, the activities of off-farm capital—specifically agrochemical companies—leave little in the way of public records, making them difficult to place precisely in

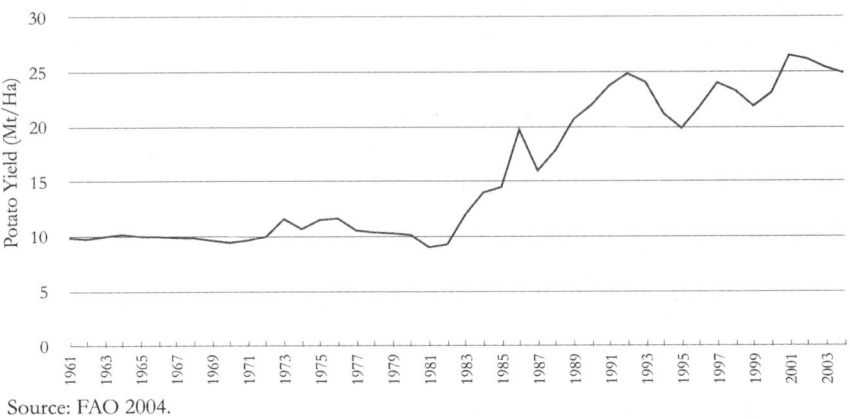

Source: FAO 2004.

Figure 3.3 Costa Rican potato yields, 1961–2004.

agroenvironmental history. It is possible, however, to consider their influence based on other research (Altieri 2000; Bull 1982; Castleman and Navarro 1987). Bull (1982, 92) notes that "the promotion in the Third World of a potentially hazardous technology requires a particularly high level of social responsibility. It requires that considerations of safety be given paramount priority even if this appears to be at the expense of sales. The reality, unfortunately, is very different. Promotional efforts often fail to give full and correct information."

Industry promotional efforts use almost all media, and advice is rarely consistent with sensible pest control that uses agrochemicals as a last resort, such as that promoted by the original conceptualization of integrated pest management (IPM). Advertising often offers impossible guarantees of good harvests and high profits and presents pesticides as a panacea for pest problems. "The message is that, if you have a pest problem, a chemical is the solution" (Bull 1982, 93) and that even if you don't have a pest problem, you should spray anyway to be on the safe side. Bull's book provides many examples of advertisements for banned products, misleading safety claims, impossible guarantees, and calendar spraying recommendations in Latin America and other parts of the world.

In the study site, agrochemical advertisements are prominent on the landscape and in agrochemical stores, as has likely been the case for decades. The billboard located just north of Cuesta La Chinchilla on a major roadway features the major vegetable crops of the area, gives a list of the company's agrochemical products, and makes the impossible claim "Your crop 100 percent protected" (figure 3.4). Promotional materials are everywhere in the area, including pesticide logos on baseball hats, bumper stickers, and even spare tire covers. In one agrochemical store, agrochemical advertisements featured scantily clad women, perhaps a new low in the use of sexual imagery to sell commodities. One pesticide salesperson in the study site mentioned that pesticide salespeople compete for prizes such as bicycles and refrigerators based on sales numbers. This type of incentive program is "outlawed" by the voluntary International Code of Conduct of the Food and Agriculture Organization of the United Nations that governs agrochemical sales internationally (Bartlett and Bijlmakers 2003), but this matters little when profits are on the line and enforcement is absent.

Despite lacking specific information on agrochemical sales promotion, in most agricultural industrialization efforts there is a two-pronged

Figure 3.4   False promises in pesticide advertising. This billboard is near Cuesta La Chinchilla, Northern Cartago. (Source: author)

strategy involving off-farm capital and the state. The potato technification campaign likely involved extension agents from the Ministry of
Agriculture and Livestock (MAG) and agrochemical salespeople visiting
farmers to convince them to adopt new seeds and more agrochemical
technology to protect them. Even though MAG greatly scaled back its extension efforts for national market crops under structural adjustment in
the 1980s (Conroy, Murray, and Rosset 1996; Edelman 1999), off-farm
capital's promotional eforts continue, and the availability of alternative
pest control information is almost nonexistent by comparison. Whether
or not state extension agents ever served as a balance to capital's agrochemical promotion, currently farmers are left almost entirely to depend
upon agrochemical salespeople for their farming information.

From Truck Farming to Export Agriculture: Chayote and Minivegetables Take Off

By the start of export agriculture in the area in the mid-1970s, intensive
agrochemical use in national market vegetables was the norm, spurred on
by state programs, the agrochemical industry, and the processes of ecological change and agrarian capitalism unfolding in the area, especially
the adoption of high-value crops. It was within this context of pesticide-
intensive national market vegetable production that vegetable production
for export began, starting with chayote in the Ujarrás Valley in the 1970s
and minivegetables in Northern Cartago in the mid-1980s. I divide this
transformation into two distinct periods dominated by different logics in
relation to pesticide use. In the first, from the 1970s until the mid-1980s,
agrochemical use on the major export crop—chayote—likely increased
relative to national market production. In the second period, starting in
the mid-1980s and continuing to the present, pesticide-use rationalization became the dominant strategy for the export sectors.

The landscape change from truck farming to export production was
unlike the shift from subsistence production to truck farming in two
major ways. First, it was far from a complete transformation. While national market vegetables almost completely replaced subsistence crops
from 1910 to 1970, the shift to export markets was not complete, and
national market vegetables remain very important today (see chapter 1).

Indeed, the potential profits to be made with national market vegetables means that they will likely persist. Second, the type of crop did not change—vegetables were exchanged for vegetables—nor did the aim of production change, since in both production systems the vegetables are produced for market.

By 1974, chayote production was the main agricultural activity in the Ujarrás Valley (Gamboa and Serrano 2003, 89), but, as shown above, the historical record is slim on details. According to some authors, many farmers in the Ujarrás Valley converted from sugarcane and coffee production to chayote in the 1970s (Bolaños et al. 1993; Calderón Mata 1980). Export chayote in the Ujarrás Valley is currently the main crop in terms of area but coexists with many other vegetables, especially tomato, sweet peppers, green bean, and celery, for the national market as well as small areas of sugarcane and coffee, which provision the national and export markets.

The change in farmers' pesticide use following their integration into the chayote export market was likely variable, depending on whether the new chayote export farmers were previously national market chayote farmers, truck farmers focused on tomatoes and other pesticide-intensive vegetables, coffee producers, or sugarcane growers. In the early 1970s, periodic insecticide use and less frequent fungicide use appears to have been the norm for chayote (Pineda Cabrales 1973). For coffee and sugarcane farmers, a change to export chayote most likely meant an increase in pesticide use, since coffee and sugarcane are not frequently or heavily sprayed crops relative to vegetables (Castillo, de la Cruz, and Ruepert 1997; Chaverri 1999). For farmers producing tomato for the national market, a change to chayote production likely meant a reduction in pesticide use, since chayote is much more resistant to pathogens than tomato, which faces the threat of late blight. For national market chayote farmers, producing for the export market probably meant little change initially in their pesticide use because they were producing the same crop, although changed quality standards might have encouraged greater use.

Few details are available on chayote farmers' pesticide use during the early export years. Calderón Mata's (1980) survey of fifty chayote farmers in Ujarrás Valley could have allowed for a comparison of practices of chayote farmers aimed at different markets, but this comparison is not done. His research showed that 64 percent of farmers planted only the *criollo* variety of chayote (for national market), while 20 percent planted only *quelite* (the export variety) and 16 percent planted both varieties (Calderón Mata 1980, 46). The majority of chayote farmers in the sample were oriented exclusively to national market production in 1980. While Calderón explored farmers' pesticide use, he did not note any differences between farmers producing for the different markets. All fifty chayote farmers interviewed used insecticides (almost exclusively carbamates and organophosphates), 78 percent used fungicides, half used nematicides, and 48 percent used herbicides. Applications were quite frequent. Intervals of every eight, fifteen, twenty-two, and thirty days were used by 12, 36, 36, and 16 percent of farmers, respectively (Calderón Mata 1980, 55). This is not as frequent as pesticide use with potato at the time, but it is still a high level of pesticide use.

When turning to export, chayote farmers likely increased their pesticide use. For one, with a greater focus on exporting, chayote yields increased 43.6 percent between 1972 and 1983, from 78.9 kg/ha to 113.2 kg/ha (Banco Central de Costa Rica, cited in Zúñiga, González M., and Fonseca Z. 1986, 172). It is likely that this increase occurred through heavier pesticide use and/or fertilizer use, but it could also have resulted from changed cultural practices or varietal improvements. More important, quality norms for export were very strict and incentivized farmers to spray more (Calderón Mata 1980). Some diseases, especially *vejiga* from the fungal pathogens *Mycovellosiella cucurbiticola* and *M. lantana* (Vargas 1988, 1991), presented considerable problems for long-distance shipments. Exporters rejected boxes of chayote showing the small discolored water-filled bulges on the fruit's surface because the disease could spread to infect the rest of the shipment. From a survey of twenty-nine chayote farmers, Zúñiga (1986) showed a significant negative relationship between the value of chayote produced for the national market and the total value of fungicides applied and showed a positive but not significant relationship between the value of export chayote produced and the total value of fungicides used. He suggests that higher levels of fungicide use decreases the damages caused by diseases, increasing fruit quality and thereby decreasing the proportion of a farmer's harvest going to the national market and increasing the proportion exported. This suggests a connection between pesticide use and fruit quality: as fungicide use goes up, the proportion of the crop meeting export quality standards increases. Thus, export chayote farmers in the 1980s might have used fungicides more heavily than those producing exclusively for the national market.

However, this singular pressure to increase pesticide use did not last. Events soon shifted the pesticide-use regime to the second period identified above, toward rationalization of pesticide use on export vegetables. About a decade into chayote production for export, the chayote export sector hit a snag—the U.S. Food and Drug Administration (FDA) detected illegal pesticide residues, leading to a shipment rejection that caused $8,000 in losses for the exporter. This occurred in 1985 when FDA chemists found chayote with residues of methamidophos, an organophosphate insecticide featured in later chapters (Thrupp, Bergeron, and Waters 1995, 155–57). In the mid-1980s, the FDA increased its oversight of pesticide residues on produce imported into the United States, which sparked years of crisis for agroexports from Guatemala and the Dominican Republic (Conroy, Murray, and Rosset 1996). In contrast, the chayote sector in Costa Rica responded quickly. After the first rejection due to illegal residues, chayote farmers held meetings about the "problems of pesticide residues in the fruit" (Valverde G. 1986, 198). The substantial financial loss for exporters and the possibility of more in the future pressured exporters and farmers into better understanding the residue problem and changing production practices to comply with U.S. regulation.

The limited information available suggests some success in controlling chayote farmers' pesticide use. In a 1980 survey of chayote farmers, 78 percent used fungicides, while 100 percent used insecticides (Calderón Mata 1980). A 1986 survey of chayote farmers, conducted after the methamidophos residue incident, shows that 100 percent of farmers used fungicides, while only 65 percent used insecticides (Zúñiga, González M., and Fonseca Z. 1986). Since an insecticide, methamidophos, caused the problem, the difference could be that farmers decreased their use as a result. In contrast, fungicides had not caused the rejections, and *peca blanca,* a fungal disease caused by *Ascochyta phaseolorum,* appeared in the early 1980s and by 1983 became chayote's most problematic fungal pathogen (Bolaños et al. 1993, 223).

Despite concern leading to action in the export sector and fewer farmers using insecticides, the problem of illegal residues in chayote did not go away. A 1987 FDA report notes that one Costa Rican chayote exporter had been subject to "automatic detention" in which it had to demonstrate compliance with U.S. residue standards before shipments were permitted entry (Food and Drug Administration Pesticide Program 1988, 164A). These costly automatic detentions continued into 1988 (Food and Drug Administration Pesticide Program 1989) and perhaps later, although this is impossible to determine since the FDA stopped publishing its list of automatic detentions after 1989. Exporters and eventually the state increased efforts to prevent farmers' use of methamidophos and related pesticides that caused rejections (see chapter 4).

Unlike chayote, minivegetables for export have been produced in the area for less than thirty years, being introduced in the mid-1980s as incentivized by Costa Rica's structural adjustment-imposed "Agriculture of Change." The minisquashes, including scallop squash and zucchinis (a major focus of chapter 2), are the most important minivegetables. The other important export minivegetables are green beans and carrots, harvested when very small, and sweet corn, which is harvested at regular size but typically broken into three- to four-inch lengths for packaging and sale. The export firms also contract with farmers to produce many other specialty vegetables mostly for the high-end national market, including artichoke, cherry tomato, Chinese cabbage, eggplant, endive, fennel, jicama, kohlrabi, lettuce, radicchio, romanesco, spinach, and tomatillo.

Those who originally introduced minivegetables began with the idea of producing them with low and rational agrochemical use (Breslin 1996). This was for two main reasons. First, the new exporters were aware of pesticide residue problems faced by export sectors in other countries, especially Guatemala. Second, one exporter's manager was formerly an organic farmer in the United States, and when he first brought the minivegetable seeds to farmers in Northern Cartago,[8] he tried to convince farmers to grow them organically. He noted that farmers made fun of him because he was very ignorant, given their experiences with vegetable production in the area. He had to change this expectation due to the high pest pressure in the area: "So, we adjusted to the minimum use [of pesticides]. . . . I had to get rid of my dogma and deal with the minimal use, and trying to educate and inform them that [pesticides are] really dangerous stuff for everybody; it has residues all the way down the line" (Mini-Horta Manager, interview, June 2000).

In contrast to the early years of chayote exports, minivegetable production required an almost immediate decrease in pesticide use for the farmers involved. Farmers switched to the minivegetables from very pesticide-intensive crops, especially potato and carrot. Minivegetable farmers became more cautious about their pesticide use as it relates to pesticide residues. One key part of this caution was a general reduction in their pesticide use, accomplished through better spraying practices and the adoption of organic production methods and/or IPM. Farmer Roberto Jiménez noted that despite exporters' demands for less pesticide use, some farmers initially snuck out at midnight to spray their crops out of fear of pest damage: "This eventually led to some of the original members leaving the group. It was a real cultural clash. They didn't want to be programmed, to be told when they could do certain things and when they couldn't" (Breslin 1996, 33). Policing farmers' pesticide use is a significant challenge for export firms, as examined in chapter 4.

The Lessons of Agroenvironmental History

While political ecologists usually study environmental degradation in an obvious material form, such as soil loss and deforestation, invisible environmental risks and hazards from environmental contamination are also important. As noted in the Introduction, critical social scientists have argued that export market engagement leads to increased pesticide use and environmental degradation, which is a specific instance of political ecology's degradation and marginalization thesis. This story does not fit Northern Cartago and the Ujarrás Valley in part because of its specific agroenvironmental history.

Vegetables produced for national market, especially potato, became pesticide intensive in the 1950s and then became increasingly pesticide dependent over the decades. It was endogenous agrarian capitalism that developed through the integration into the national market (truck farming), not a shift to export orientation, that created the political-economic conditions favoring heavy pesticide dependence. Instead of increasing pesticide use from already high levels, national market vegetable farmers who adopted new export crops in Northern Cartago and the Ujarrás Valley faced pressures to rationalize and reduce their pesticide use to comply with what they came to view as strict market requirements for low or no residues, as we will see in chapter 4.

These findings demonstrate the problems with the scalar trap (Brown and Purcell 2005), in which political ecologists and others in allied fields assume that local production and political control of resources results in more sustainable human-environment relations than organization at other scales. Implicit is that integration into local and national markets is less harmful to local environments than integration into the export market. This overlooks the spread of agrarian capitalism through national market integration or assumes that nationally and locally organized agrarian capitalism is somehow a kinder, gentler form of agrarian capitalism. Marx's discussion of capitalist agriculture undermining the conditions of production (see chapter 1) does not distinguish at all between market scales but instead uses the development of English agriculture as his case. It is the spread of agrarian capitalism—based on extractive, competitive, grow-or-die, short-term profit maximization logic and the subordination of producers to off-farm capital—that undermines ecological sustainability. Before society reins in capitalist agriculture's most egregious problems in a Polanyian double movement, agrarian capitalism, regardless of how near or distant its markets are, has detrimental environmental and social effects locally, especially when compared to peasant modes of production based mostly on use value and from which little is extracted.

Thus, we must recognize that environmental degradation can be created through the expansion of local and national markets as well as international markets. National market expansion and regulation are important but severely understudied processes in most developing countries. For scholars in the North, national market expansion remains largely unexamined because of thinking that remains constrained by the scalar trap. Many other topics of interest to political ecologists and environmental historians could be addressed through symmetrical comparisons of the environmental effects of export and national markets. Indeed, a comparative understanding of the logics of various capitals operating in these markets and the ways in which they are constrained and enabled by the state would reveal a great deal about the sustainability of various configurations. These comparisons will become all the more interesting with social movements' and consumers groups' continued demand for increasingly regulated agrifood markets in industrialized countries and the often incipient expressions of these pressures in developing countries.

The geographical significance of the history told in this chapter is that the narrative of increased pesticide use with new export production in the 1980s—which has previously been considered to apply in a blanket fashion to Latin America—is actually geographically variable and depends on prior agricultural systems. Thus, paying attention to local agroenvironmental history is essential for understanding the environmental impacts of new agroexports, since environmentally benign national or local market production cannot be assumed to exist prior to export production.

This agroenvironmental history also creates an interesting perspective from which to view calls for more local and national provisioning, often called food sovereignty, a commonly advanced solution from agroexport critics. Pursuing more nationally based provisioning is not without problems.[9] Paying attention to agrarian capitalism means recognizing that national market expansion is often at odds with how food sovereignty promoters (and export market detractors) conceptualize it. Potato production in the study site illustrates some of these difficulties. Costa Rica has long had protective tariffs on potatoes—these tariffs even survived structural adjustment (Edelman 1999)—and still retains a 46 percent tariff on imported potatoes even after entering into recent free trade agreements (CentralAmericaData 2010). The Central American–Dominican Republic–United States Free Trade Agreement (CAFTA-DR), negotiated in 2004, eliminated tariffs on all imported foods into Costa Rica except for fresh potatoes and fresh onions (Foreign Agricultural Service 2005a), major crops of Northern Cartago. CAFTA-DR met considerable protests and organizing, although the agreement eventually passed by a narrow margin in a popular vote in Costa Rica, going into effect in January 2009 (Cupples and Larios 2010). All other signers of CAFTA-DR eliminated potato tariffs over fifteen years, while Costa Rica only agreed to slowly increase its quota for U.S. potatoes (Foreign Agricultural

Service 2005a).[10] Similarly, Costa Rica's tariffs on fresh potato imports were excluded from the Canada–Costa Rica Free Trade Agreement, allowing them to remain intact (Agriculture and Agri-Food Canada 2012).

While Lizano (1994) argues that Costa Rican potato production is relatively efficient, a geographical analysis of production (see chapter 2) shows that if subjected to market liberalization, the losers in these heavily pesticide-dependent production systems would likely be those farmers who spray the most. These tend to be the more resource-poor farmers in the cloud belt facing the most intense pest pressure. Because of thinking limited by the scalar trap, these kinds of difficult environmental and social consequences of national provisioning do not receive much critical attention. Certainly some production problems can be reworked with agroecological methods and approaches, as suggested by food sovereignty promoters (Perfecto, Vandermeer, and Wright 2009), but producing potatoes in Northern Cartago without synthetic inputs is currently a virtual impossibility as farmers see it.

Costa Rica, then, currently faces a difficult food sovereignty trade-off with potato tariffs. Maintaining its tariffs—a position in line with nationally maintained trade decisions that can protect domestic producers—helps sustain farmers' livelihoods in the regions that are least well suited for potato production, which means that extremely high levels of pesticide use remain. Considering the workings of agrarian capitalism, the trade protections also serve to protect input makers' and input retailers' markets, as potato farmers in the lowest elevations spend upwards of $2,500 on agrochemicals per acre during a crop cycle. When applied over hundreds or thousands of acres, these input sales amount to hefty incomes and profits for off-farm capital.

Recognizing this, the food sovereignty questions are: Can Costa Rica maintain the protections on these important vegetables and move them away from being the most agrochemically intensive on the planet? Is there political will to make this possible? With small- and medium-scale farmers protesting to lower pesticide prices (Anonymous 2008a), food sovereignty proponents' calls for national provisioning based on agroecological methods will take considerable efforts and must confront the workings of agrarian capitalism in the country, including farmers' complete dependence on pesticides for their livelihoods. Costa Rica did pass an organic standard in 2006 (Aistara 2011), and in 2010 a group called Paren de Fumigar (Stop Spraying) formed to contest Costa Rica's heavy pesticide dependence and advocate for the prohibition of red-labeled pesticides (those most acutely toxic). Yet there is a long way to go before Costa Rica as a nation turns away from heavy dependence on pesticides and embraces alternative agriculture in line with its green republic reputation, an issue that I take up again in the Conclusion.

Overall, detailed agroenvironmental history as advocated by environmental historians (Crosby 1990; Worster 1990) has an important role in political ecology. In this case, agroenvironmental history reveals that the common narrative about increased pesticide use from new agroexports in Latin America does not apply. Exporters soon became concerned about violations caused by residues and thereby began policing farmers' pesticide use. This raised its own challenges, to which we now turn.

4 Policing Pesticides

Ricky Montaldo, a vegetable export firm manager, leads me into the small room packed with papers that serves as his office. He has kindly agreed to talk with me and is interested in my research in part because little precise information about U.S. regulation is readily available to him. I sit across from him as he sits at his desk. An old but still functional computer lurks behind him, and posters of varieties of vegetables adorn the walls. The one office window looks out over the produce-handling floor, where young women in hairnets make farmers' vegetables more palatable to an outside world by cleaning, drying, and neatly packaging them. Ricky chain smokes, and his ashtray overflows.

Vegetable export firms occupy the invisible but crucial middle of the food system. Ricky's is a high-stress job, involving tight coordination of incoming produce from farmers with market orders in the United States and Canada and for high-end national markets. Through contracts with farmers, he coordinates the production of dozens of vegetables—considering quantities and various qualities simultaneously—for many markets on multiple time scales: the minute, hour, day, week, month, and growing season. Every few minutes during our conversations the phone rings, and Ricky excuses himself to answer it. Most of the time things work out, but sometimes they don't and produce isn't moved. As Freidberg (2009, 195) notes, "everyone in the business has at some point watched freshness die."

Ricky tells me about issues that he faces, how he understands U.S. food regulations, which farmers he buys vegetables from, and how he tries to shape their production practices to comply with regulations. Pesticides are a major concern of his because other export firms have experienced rejections due to illegal residues and because a rejection is a direct economic loss, since the shipment is destroyed and the importing produce buyer does not pay him. A rejection also negatively affects his relationship with produce importers in North America, since they are less likely to trust him to deliver. Because of this risk, he must contract with farmers he can trust, and he communicates which pesticides they can and cannot use. His side of the deal is to maintain a fixed price for the produce he buys, a rarity in the world of Costa Rican vegetables, where prices fluctuate dramatically over short periods of time. The fixed price matters a great deal to farmers, as discussed in chapter 6, because they are able to make decisions on inputs that make sense economically and that comply with Ricky's terms.

This chapter highlights how food industry firms in the study site adapt to regulation through their contract farming relationships. I focus on two vegetable export sectors in Costa Rica, chayote and minivegetables, the histories of which were described briefly in chapter 3. I examine how exporters understand U.S. market requirements and how their contracts with farmers work in terms of communicating pesticide residue regulations, exercising various forms of control, and promoting pest control alternatives, if at all. While much literature on contract farming notes that produce buyers exercise considerable control over farmers, the extent of off-farm capital's control over farm families' labor process cannot be predicted with certainty (Clapp 1994; Little 1994) but is instead a topic for further empirical research.

Pesticide Residue Regulations and Contract Farming

Pesticide residues are one of the ghosts that haunt our modern age. Despite pesticide residues seeming like a uniquely modern concern, historian James Whorton (1974) has documented concern over pesticide residue in the United States long before Rachel Carson's *Silent Spring*. A period of heightened concern followed the publication in 1906 of Upton Sinclair's *The Jungle* (Sinclair 1985) and helped to pass the Food and Drug Act of 1906, the first major food regulation act in U.S. history. This act established oversight of contaminants in the food supply (Whorton 1974).

Yet tolerances for agrochemicals, also called maximum residue levels (MRLs), were first established in the United Kingdom before *The Jungle* was written and had important trade implications. While "tolerance" suggests to some people that these levels are those that can be tolerated by humans, these are tolerances in the sense of standards: "They are literally the limits of what behavior will be tolerated . . . without incurring some sort of negative sanction" (Busch 2011, 25). The first pesticide tolerances for food, regulating the formerly popular arsenical pesticides, resulted from the epidemic of alcohol peripheral neuritis that occurred in Manchester, England, in 1900. The epidemic killed seventy people, caused six thousand illnesses, and created substantial concern about toxins in food and drink (Whorton 1974, 84). The majority of those affected were industrial workers who drank inexpensive beer, which led investigators to find that the poisoning resulted from arsenic-contaminated sulfuric acid used to make glucose, a cheap malt substitute in beer making. A 1901 UK commission dealing with the problem recommended that arsenic—regardless of its source, be it pesticides or contaminated food additives—be limited to 0.01 grain (0.65 milligrams) per pound of solid food (Whorton 1974, 87). The British government informally adopted the standard, and it became a de facto world tolerance for arsenic. This first tolerance directly impacted the fledgling transnational fresh produce trade since enforcement in the United Kingdom meant that U.S. produce shipments, especially apples, often violated the tolerance, and the United Kingdom threatened an embargo.

Governments continue to regulate pesticide residue levels in food. They do so because of uncertainty about the negative effects of residues, consumer concern about ingesting pesticides, and industry's liability concerns. There are also agreed-upon international standards in the Food and Agriculture Organization's *Codex Alimentarius* (Food and Agriculture Organization 2008), yet these are nonbinding and not commonly used by industrialized nations. These governments tend to set their own pesticide tolerances, often using scientific risk assessments based on rodent laboratory tests (Abelson 1994; National Research Council 1983, 1987). In the United States today, tolerances are generally set in the range of 0.1 to 50 parts per million (ppm, equivalent to mg/kg), and detection levels of specific pesticides are generally around 0.01 ppm but range between 0.005 and 1 ppm (Food and Drug Administration Pesticide Program 2004, 3). Once the Environmental Protection Agency (EPA) sets a tolerance, the Food and Drug Administration (FDA) is responsible for testing the crop-based portion—as opposed to animal-based portion—of the U.S. food supply, including imports, for pesticide residues.

Since pesticide-use patterns are not often deeply studied and understood in most places, the FDA employs multiresidue methods (MRMs) that screen at one time for 100 to 200 pesticides of the many hundreds that could appear as residues. Because of the expense of pesticide residue testing, most foods in the United States are tested infrequently. For example, of the approximately 10 billion bananas imported into the United States every year, an average of 167 are sampled (Wargo 1998, 158). Fresh vegetables imported into the U.S. from its three largest suppliers—Mexico, Canada, and Costa Rica—are subjected to 1 pesticide residue test per 8 million, 2.5 million, and 2 million kilograms (kg) imported, respectively (Galt 2010).

Many critics of the EPA's pesticide tolerance setting and the FDA's pesticide residue testing argue that their efforts fail to keep the U.S. food supply safe for all members of the population (Galt 2009; General Accounting Office 1979; Wargo 1998). Yet despite its many inadequacies, the FDA Pesticide Program's residue monitoring has real effects on export commodity chains, since they are standards that are tied to sanctions. U.S. residue monitoring has caused serious problems and economic losses for some countries and exporters that were not in compliance, especially Guatemala and the Dominican Republic in the 1990s (Conroy, Murray, and Rosset 1996; Murray and Hoppin 1992) and Guatemala, Spain, and China in the 2000s (Galt 2010). Thus, pesticide tolerances are important in a political economic sense, regardless of whether they are effective in protecting human health, because they are passed along as market requirements to the various actors involved in export commodity chains. This monitoring creates a risk of shipment rejections, as Ricky Montaldo and other export firm managers were quick to note.

Examples of regulation from afar impacting distant production systems abound, including the United Kingdom's arsenic tolerance discussed above. Costa Rica's DDT ban is another. The United States threatened to halt all Costa Rican beef imports because of high DDT residue levels in the 1980s, and local authorities quickly banned DDT importation and use (Anonymous 1988a, 10). The threat of large losses to foreign exchange earnings and the significant power of Costa Rican beef producers

(Edelman 1995) who had no use for DDT drove the banning of DDT's agricultural use in Costa Rica. Similarly, aldicarb—a very acutely toxic carbamate insecticide and nematicide—was withdrawn from use in Costa Rica in 1991 (Dinham 1993; Matute Ch. 1991; Wheat 1996). Prior to the ban, it was clear that there was a high rate of aldicarb poisoning of workers in banana plantations, which often resulted in deaths. This was not the impetus for the ban, however. It "was due . . . to the danger it posed for the consumers in the US after residues over the tolerance limits were found in bananas shipped from Costa Rica and other countries" (Dinham 1993, 103).

More generally, by 1990 Costa Rica had banned twenty-four of the twenty-six pesticides whose registrations the EPA had canceled and/or suspended (General Accounting Office 1990, 27). These bans illustrate the unequal distribution of power—in the form of the ability to influence policy outcomes—in a world tied together through trade in fresh produce. These pesticide residue risks to consumers in the industrial world have led to significant changes in pesticide use in export production systems in the developing world, while the death and suffering of local populations generally has little impact on policy.

When shipping fresh produce to the U.S. market, exporters need to minimally fulfill three often contradictory market requirements: conforming to U.S. phytosanitary standards that require a complete absence of pests and pathogens; meeting high cosmetic standards for unblemished produce (Pimentel, Kirby, and Shroff 1993; Trivelato and Wesseling 1992); and complying with EPA pesticide tolerances. When exporters pass these market requirements on through their contracts, farmers find themselves in a difficult situation in which they are expected to, on the one hand, potentially decrease pesticide use to comply with tolerances and, on the other hand, use pesticides frequently in high volumes to meet strict phytosanitary and cosmetic requirements (Thrupp, Bergeron, and Waters 1995). Despite neoliberalization's rapid spread in most parts of the world, pesticide residue regulations remain intact (Galt 2011). Within World Trade Organization (WTO) rules, "the defense of consumer health and safety became one of the few permissible 'non-tariff trade barriers,' and one that wealthy WTO member states have invoked repeatedly" (Freidberg 2004, 9).

How do these contradictory pressures play out in commodity chains and in land-user decisions about pesticide use? Social scientists and agroecologists studying Central American agriculture in the 1980s and 1990s argued that pesticide problems were the Achilles' heal of agroexport promotion (Rosset 1991) and pointed to pesticide residue problems as a sign of farmers being caught on the pesticide treadmill (Murray 1994). Critiques of new agroexports in the 1990s were especially strong, yet social scientists and agroecologists have largely dropped the topic to move on to alternative agrifood movements and governance, especially fair trade and organic certification. They left behind what they had argued were export sectors in social and environmental crisis. As a result, there are no studies of possible adaptations of Latin American export firms to food system regulation.

Rather than succumb to the regulatory problems created by pesticide residues on their produce and the pesticide treadmill, actors in some export sectors adapted to regulatory challenges by modifying social and biophysical aspects of their production systems. In contrast to agrochemical manufacturing and sales firms, which are off-farm capital with little interest in the specificities of the production system as long as pesticide sales are occurring and increasing (see chapters 2 and 3), exporters as off-farm capital have a direct economic stake in the performance of the commodity chain. In other words, while the pesticide industry can push and promote pesticide use, produce firms face economic losses from regulatory noncompliance. Most therefore aim to shape the production system in what are often called coordinated supply chains, which involves struggles over farmers' production practices.

While analyses in the 1990s tended to emphasize the widespread nature of the ecological and regulatory crisis of new agroexports, more recent analysis reveals substantial variations in commodity chains' compliance with U.S. regulations (Galt 2010). Table 4.1 shows the percentage of vegetable samples containing illegal residues, as tested by the FDA between 1996 and 2006. Some countries, such as Turkey, Italy, Peru, and Chile, have very low rejection rates, while others—Guatemala, Spain, Jamaica, and China—have high violation rates. Costa Rica, the third-largest vegetable exporter to the United States after the giant exporters Mexico and Canada (Galt 2010), competes with fifty-six other countries for U.S. vegetable markets, and the ability to pass through FDA inspection without rejections is an important factor in remaining competitive (Julian, Sullivan, and Sánchez 2000). Costa Rica had a rejection rate of 4.4 percent due to illegal residues (see table 4.1), lower than the average of vegetable imports into the United States (5.2 percent) and slightly lower than Mexico's (4.6 percent). Overall, Costa Rica lies on the low to middle range of compliance, with many countries doing better and many doing worse, some considerably. This is a point of pride, and Costa Rican newspapers will sometimes report a year of success of exports without rejections (Anonymous 2008b). These comparisons suggest that some countries and the export sectors grounded in them have taken U.S. regulations seriously and adapted, while others have not done so.

The rest of this chapter focuses on the efforts of the five major export firms in the study site.[1] Two export firms, Mini-Horta and HortaRica, specialize in minivegetables and originated as one firm that splintered because of disagreements. The three chayote export firms included in the study—ChayoRico, ChayotEx, and SuperChayote—are the largest in the area.

These locally owned export firms contract with farmers to provide their produce. I use an encompassing definition of contract farming: "Those contractual arrangements between farmers and other firms, whether oral or written, specifying one or more conditions of production and marketing of an agricultural product" (Roy 1927, 3, cited in Watts 1992, 69). Complete contracts spell out all obligations of both parties in every eventuality. These are rare in developing countries and in the study site, where exporters instead use incomplete contracts that are often verbal. As a social arrangement, contract farming "presupposes some form of regulation and control, a sort of direct fashioning, of the labor process by the contractor, and a web of social relations, which are practically and ideologically central to the production system" (Watts 1992, 70). In sectors where export firms are not in direct control of the production process, as is the case for many vegetable sectors, it is essential to examine contract farming to understand commodity chains' adaptions to regulation. Below I focus on their pesticide policing within these contract relationships.

Table 4.1 Pesticide residue violation rates of vegetables tested in the United States by country, 1996–2006

North America		Europe	
United States	1.6%	Belgium	1.3%
Canada	1.9%	Bulgaria	0%
Latin America		France	5.8%
Belize	10.0%	Germany	0%
Brazil	0%	Greece	7.8%
Chile	1.9%	Hungary	0%
Colombia	3.1%	Italy	0%
Costa Rica	4.4%	Netherlands	1.1%
Dominican Republic	7.8%	Norway	0%
Ecuador	3.8%	Poland	3.7%
El Salvador	1.6%	Portugal	0.0%
Guatemala	18.3%	Spain	17.5%
Guyana	0%	United Kingdom	9.1%
Honduras	7.5%	**Southwest Asia & North Africa**	
Jamaica	13.6%	Egypt	8.6%
Mexico	4.6%	Iran	0%
Nicaragua	4.3%	Israel	2.2%
Panama	0%	Lebanon	3.2%
Peru	1.9%	Syria	0%
Trinidad and Tobago	8.2%	Turkey	0%
Venezuela[a]	20.0%	United Arab Emirates	0%
East and Southeast Asia		Oceania	
China	13.2%	Australia	6.3%
Hong Kong	12.5%	Fiji	7.1%
India	4.1%	New Zealand	0%
Indonesia	0%	Tonga	0%
Japan	0%	Western Samoa	0%
Philippines	4.8%	**Africa South of the Sahara**	
South Korea	8.2%	Cameroon	0%
Thailand	1.1%	Ghana	0%
Vietnam	6.1%	Ivory Coast	0%
		South Africa	0%
All imports	5.2%		

[a] The sample-size is very small; one of five tests showed residue violations.
Source: Analysis of Food and Drug Administration (2013), PROD database 1996–2006.

Policing Pesticide Use

To address questions of how pesticide regulations are socially mediated by exporters, I interviewed the managers of the five major exporting firms in the minivegetable and chayote export sectors of Costa Rica, located in Northern Cartago and the Ujarrás Valley, respectively. All emphasized the importance of export markets' pesticide tolerances and the negative impacts of produce rejections caused by illegal residues. They made it clear that past residue problems in Costa Rica and other countries have made pesticide policing—control over the specific pesticides that farmers use and how they use them—a necessity. One chayote exporter explained that this is because of the financial losses involved, between $10,000 and $15,000 per shipment with illegal residues. "It is not an everyday problem. But it happens, and it is a serious problem" (ChayoRico manager, interview, November 2003).

These interviews occurred around the time that exporters had to start complying with the requirements of the U.S. Public Health Security and Bioterrorism Preparedness and Response Act of 2002, also known as the Bioterrorism Act. One of the many consequences of the events of September 11, 2001, was the realization that food and water supplies are vulnerable to bioterrorism, or "the purposeful use of biological or chemical materials to achieve political goals" (Nestle 2003, 249). The new law added requirements for overseas exporters who ship to the United States. All exporting firms must register with U.S. agencies and notify U.S. Customs in advance of each shipment. The law also requires substantial changes in practices for exporters and farmers, including more hygienic practices in packing sheds and on farms and the use of good agricultural practices (GAP), which involve methods to improve food safety by reducing chemical and microbial hazards and ostensibly promote environmental sustainability on the farm (Food and Agriculture Organization 2003). GAP promotion comes from retailers' fears of liability from potentially dangerous food products. Under GAP, farmers must maintain logbooks of agricultural practices, including agrochemical use. These logbooks can then be audited if problems occur.[2] Since the Bioterrorism Act was going into effect, exporters brought up the topic during the interview. While not its main focus, the discussion below examines some of the preliminary effects of the changes.

Export managers do not fully understand the complicated U.S. regulations regarding pesticide residues, despite their economic importance. Information on the specifics of U.S. regulation and enforcement remains unavailable to many exporters, and the fact that the United States divides pesticide residue oversight among three different agencies—the EPA, the FDA, and the U.S. Department of Agriculture (USDA)—understandably generates confusion. Nonetheless, exporters share basic understandings. All export firm managers understood that the United States, Canada, and European nations regularly check for pesticide residues on imported produce. However, the managers generally did not know details, such as which agency is responsible for the testing, how frequently testing occurs, which pesticides are tested for, which crops are the most frequent violators, and the specific reasons why violations occur (i.e., whether they result from a residue of a pesticide not registered on the crop or from a residue of a registered pesticide that is too high). This lack of knowledge is not due to a lack of interest but instead results from the facts that the FDA does not publish detailed analyses showing the most common violations caused by pesticides, crops, and pesticide-crop combinations and that the EPA does not have an accessible or complete system for communicating its pesticide tolerances (Galt 2009). It is common knowledge among export firm managers, however, that some countries, especially Guatemala, have significant problems with produce shipment rejections due to illegal residues.

Information that export managers receive concerning shipment losses and detentions due to FDA pesticide residue testing is fuzzy. This is in part because the FDA does not contact managers when a detention occurs. Instead, it is the importer in the United States that is responsible for this communication (López-García 2003). Managers have heard of or experienced shipments being rejected because the FDA refuses them entry into the U.S. food market. They also told of hearing about their produce rotting from delays due to further residue testing to verify the previous analysis. Additionally, managers noted that importers in the United States can deceive small-scale independent exporters in developing countries, telling them that the shipment was lost due to rot or residue violation when they actually received it and sold it.[3] Exporters in the study site voice suspicion that this happens, but no one has investigated the extent of these abuses.[4]

Additionally, the Costa Rican press is not always a reliable source of information on export problems. For example, *La República* (Anonymous 1988c, 2), one of Costa Rica's leading newspapers, ran a story in 1988 stating that "the United States suspended the purchase of chayote and other agricultural products from Costa Rica, given the high level of pesticides dangerous to human health detected by the Health Department of that country."[5] The story also reports that this is a large economic loss because the United States returned the shipments and suspended the purchase of Costa Rican produce. The following day, *La República* ran a story essentially retracting the previous one in which Juan José May of the Ministry of Agriculture and Livestock (MAG) told the newspaper that he had not received any official communication from the USDA about chayote or the stoppage of exports. He pointed out that for this to happen there must be three warnings and noted that chayote exporters continued to send their shipments (Anonymous 1988b). What had apparently happened was that one exporter had a shipment rejected due to pesticide residues and contacted the Ministry of Health (Ministerio de Salud), not MAG, so Costa Rican government agencies were not on the same page.

Other reports of rejections due to violative pesticide residues are incomplete. For example, *La Nación* reported that from August to October 1990, 45,000 kg of chayote and 11,358 kg of squash were rejected "for having pesticide residues" (Brenes 1991, 5A), but the story does not note that some residues are permitted and others are not. As such, exporters must act on limited and imperfect information—from importers, U.S. regulatory agencies, and the Costa Rican press—to make their produce comply with residue regulations.

Minivegetable Exporters' Forms of Control

Costa Rica's two minivegetable exporters—referred to here as Horta-Rica and Mini-Horta—are both located in Cipreses in Northern Cartago (see figure 1.1). Minivegetable exporters sell to both export markets and to the national market. For the export market, they focus mainly on minisquash and green beans. The main markets are wholesale buyers who provision restaurants and hotels in the United States and Canada, specifically Miami, Los Angeles, and Montreal. Sales to the high-end national market serve supermarket chains, hotels, and restaurants.

I asked each manager how he understood market requirements of the United States. In response, HortaRica's manager succinctly answered:

> The market requirements of the United States, to be able to export at this moment, are those that comply with the regulations that they have, including no chemical residue, no insects, no rotting products, so that it is not rejected. And after that, well, a requirement in size and form and taste. And as of December 12, 2003, we have to comply with the Bioterrorism Act that is published by the Congress of the United States. We have to register the packing plant and [follow] other recommendations. (HortaRica manager, interview, April 2003)

Rather than a discussion of pesticide tolerances, the manager simplified U.S. demands to "no chemical residue," as did the newspaper account describing the rejections of 1990 (Brenes 1991). Similarly, Mini-Horta's manager explicitly made a simplification when I asked him about what he says to his contract farmers about pesticide tolerances. He corrected me when I used the word "tolerances," saying that he tells farmers that their produce cannot have any residues: "I didn't say tolerance, I said *any* residues" (Mini-Horta manager, interview, April 2003). Managers of both HortaRica and Mini-Horta emphasize to their contract farmers that the U.S. market requires zero pesticide residues. They know that there are pesticide residue tolerances in the United States but intentionally simplify the message and make it more stringent so that farmers will strive to have zero residues on their produce.

Part of the requirements for zero residues is that minivegetable exporters specifically prohibit the use of the organophosphate insecticides, especially methamidophos and dimethoate, which have caused previous violations for Costa Rican produce from the area. Both of these insecticides are systemic, which means that they are translocated throughout the plant and kill insects that feed on it in addition to the protection of residues remaining on the surface. As an alternative to these insecticides, exporters recommend that farmers use pyrethroids, synthetic pesticides that mimic the natural insecticide pyrethrum, are not systemic, and break down much faster. Thus, "zero residues" in the minivegetable sector functions as a type of code for "use of pyrethroids instead of organo-phosphates and organochlorines." Interestingly, the manager of Mini-Horta did not discuss control over fungicides, likely because these had not caused violations in produce exported from the area when we spoke in 2003. This exporter does not prioritize fungicides in communication with farmers. In contrast, HortaRica's agronomist cautions farmers to examine the preharvest interval (PHI) on *all* agrochemicals, including fungicides.

Knowledge of and experience with past rejections from illegal residues greatly influence how exporters understand pesticide residue regulations. Upon being asked if the company had experienced rejections due to residues or other reasons, HortaRica's manager sat back, drew in a breath, and said, "Yes, sir. [The company] had a problem once with a rejection because of a new producer in the province of Alajuela and he used [prohibited] inputs, and then this happened [resides were detected] and the shipment was destroyed" (HortaRica manager, interview, April 2003). During a different interview, a contract farmer for HortaRica identified the pesticide behind the rejection as BHC, a persistent organochlorine insecticide. To his knowledge, the farmer who caused the violation had planted minisquash on an old coffee farm on which BHC (hexachlorocy-clohexane, an organochlorine insecticide also known as lindane) had been used.[6] Because of this experience, HortaRica instituted a policy that farmers must conduct pesticide residue tests for soils from agricultural land newly brought into minivegetable production, such as land previously devoted to coffee, sugarcane, or national market vegetables. Additionally, HortaRica started making recommendations about which chemicals farmers could use and has formalized these over time so that now it has the strongest policing mechanisms among exporters in the study site.

In contrast, Mini-Horta's manager reported never having experienced any problems with pesticide residues. He mentioned that just four days before talking with me, a shipment had been held for extensive residue testing in Miami, where "they do a strict examination. They put the whole thing in quarantine; it's like they were worried about the problems with Guatemalan [produce], so they took the whole thing out and started examining everything and found zero, zero, zero, zero [residues]" (Mini-Horta manager, interview, April 2003).

Despite the problems with residues, minivegetable exporters do not use written contracts but rely instead on verbal agreements. When asked why, HortaRica's manager responded, "I don't know why. Maybe it is our culture that we don't accept contracts, that we do our business respecting the things that are talked about" (HortaRica manager, interview, April 2003). In contrast, although Mini-Horta never used written contracts in the past, the company was implementing written contracts in late 2003. Prior to and during the implementation of these written contracts, the verbal agreement between Mini-Horta's manager and farmers continued to be more important than the written contracts during my fieldwork. The main reason behind the written contracts was to increase farmers' loyalty in selling their produce to Mini-Horta. The administrator of Mini-Horta explained that the firm sells the seeds to its farmers on preferential credit but that sometimes farmers sell the produce to their competitor or do not plant during the rainy season because a greater investment in agrochemicals is needed. This leads to an inconsistent level of supply. "This affects the international market, and the national one of course, because the same quantity is not available, and then one cannot satisfy demand for the product. Thus, we have export problems, right? So management wants a contract so that the farmer, dry season or rainy season, promises to bring a certain quantity here. To secure the market" (Mini-Horta administrator, interview, April 2003).

This was the main impetus behind the contracts, but the contracts also have clauses about agrochemical use and specifically about farmers keeping logbooks of pesticides used. (HortaRica also requires farmers to use logbooks, but this is part of the verbal contract.) I asked the Mini-Horta's manager about the logbook:

It is called a *registro*. . . . [W]hen they plant they put down the planting date, they say organic fertilizer, or 10-30-10 [N-P-K content], if they have to use some kind of pesticide for the soil or not [and] the dose. For every step of the entire process. . . . What I am saying is I need that *registro* on your farm, that if a market [representative] shows up from the [United] States, or an agency from Canada or Europe comes in, they'll want to know exactly what is going on especially when you're talking export product. (Mini-Horta manager, interview, April 2003)

Mini-Horta's manager sees the *registro* as increasingly necessary for securing markets in the United States, since produce buyers demand traceability and assurance of farmers' GAP use. He later emphasized that while he was looking into marketing deals with Costco and Whole Foods, the buyer's first question was always about logbooks. These buyers need Mini-Horta's farmers to maintain logbooks so Mini-Horta can obtain the necessary insurance policy to cover problems created if consumers get sick from the produce and sue the retailer (Mini-Horta manager, interview, April 2003). Most retail outlets in the United States now require this food product liability insurance. A typical $1 million policy for a product to be sold in the retail outlet typically requires a $1,000 annual premium (Holland 1998, 1).

As part of their agreement, minivegetable exporters expect their farmers to comply with GAP. GAP's stated goals are to produce safe and high-quality food and to ensure the environmental sustainability of agriculture, and practices often include integrated pest management (IPM) and soil conservation (Food and Agriculture Organization 2003). HortaRica requires that farmers comply with its GAP list, which must be posted on each farm. As posted, the requirements pertaining to pesticides include the following:

—Decrease the use of agrochemicals for pest and disease control, and monitor their populations to spray only when it is necessary.
—Respect the periods between application and harvest [PHI] to avoid the presence of chemical residues on the product.
—Conduct analysis of water [supply], soil, and agrochemical residues in the soil.
—Keep a logbook of daily agrochemical applications, and use only the agrochemicals permitted and recommended by the agronomist.

HortaRica employs an agronomist full-time to visit its farmers, which numbered forty-six during the time of fieldwork (HortaRica manager, interview, April 2003).[7] HortaRica is the only exporter in the study site that employs an agronomist to make inspections and agrochemical recommendations to farmers. He provides technical support and pesticide recommendations to farmers but also serves as an in-the-field enforcer of the exporter's requirements. The agronomist allowed me to accompany him on one of his busy days (this occurred after my farmer survey was complete, as I did not want farmers to perceive me as being connected to the exporter). He visits many farmers a day, discussing production problems with them and writing prescriptions for agrochemicals. If a farmer is absent, the agronomist reviews the farmer's collection of old pesticide containers to see what was sprayed (figure 4.1). The agronomist's visits explicitly show the power relations involved in the social mediation of export market requirements.

In addition to the agronomist, HortaRica has paid for IPM training from Universidad de Costa Rica for its core farmers. One of the more enthusiastic participating farmers showed me his IPM resources, including a binder of information on squash and green bean diseases and diagrams to estimate the incidence of disease to determine the threshold for spraying. IPM represents an encouraging development, since most farmers in the area rely solely on calendar spraying of pesticides. Yet the farmer emphasized the need for me to keep the information secret, since HortaRica and its farmer organization had paid for the materials and training from

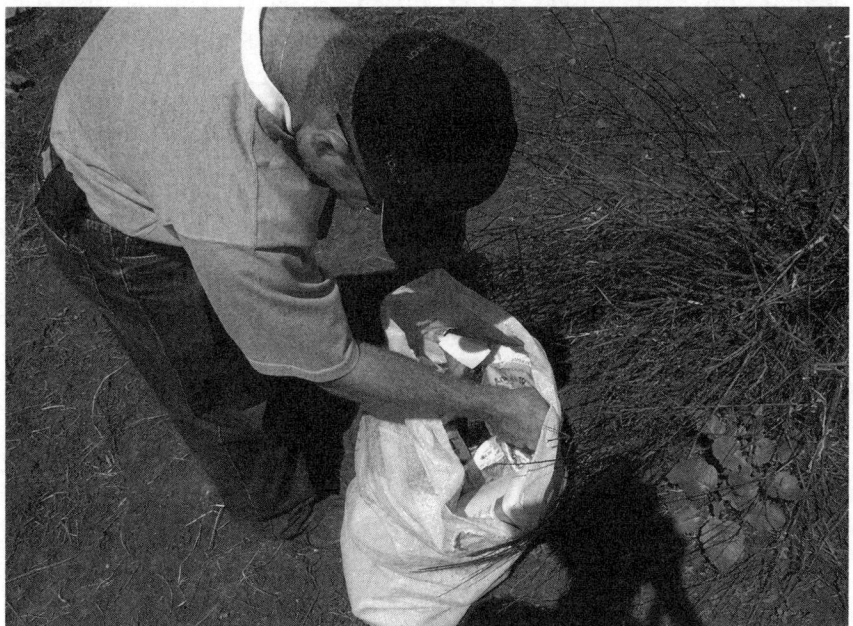

Figure 4.1   Exporter HortaRica's agronomist checks on agrochemical use. Here he is using empty pesticide containers as evidence. (Source: author)

the university and saw it as a way of gaining a competitive edge over Mini-Horta. Thus, a method that can potentially reduce farmers' pesticide use in the area if widely adopted is being hoarded for competitive advantage.

In contrast to the more resource-intensive efforts of HortaRica described above, Mini-Horta exercises markedly different types of pesticide policing, but controlling farmers' pesticide use is also a high priority. Mini-Horta does not have an agronomist. Instead, the firm usually purchases from only trusted farmers who are well known by the management team. Mini-Horta's manager encourages his contract farmers to adopt organic agricultural methods and encourages them to attend free classes on organic agriculture at the local organic agriculture school, Unidad Tecnológica en Agricultura Orgánica, run by the Instituto Nacional de Aprendizaje (National Institute of Learning) near Cuesta La Chinchilla in Northern Cartago.

For enforcement, Mini-Horta relies on the social bonds formed between farmers in an associated farmer organization. Mini-Horta helped establish this farmer organization, and the manager explained its effectiveness in controlling pesticide use. Peer pressure works as a form of within-class discipline in which farmers oversee each other. Using proscribed pesticides such as methamidophos and dimethoate "could jeopardize everybody else in the organization. So if you, as a farmer, don't listen to that and aren't conscious, don't even think about it, because you could mess up your friend and everybody else's life. Not just yours and mine, but everybody else's. And so that social consciousness really was the glue" (Mini-Horta manager, interview, April 2003).

Similarly, HortaRica also helped create export farmer organizations consisting of its contract farmers. As with group-certified organic coffee farmer cooperatives (Mutersbaugh 2002), farmers in these export farmer organizations take interest in the practices of their fellow members. I attended a field day in which farmers in one organization visited one another's farms. When we arrived at a farm with no farmer present, they headed for the sacks of empty pesticide containers. Since one's actions can negatively influence the economic future of everyone in the group, this can act as a strong disincentive for irresponsible pesticide use that could cause pesticide residue violations.

In addition to these forms of control, both minivegetable exporters reported that they have implemented traceability systems as required by the U.S. Bioterrorism Act. A traceability system maintains information on the provider of each carton in an export shipment, usually by assigning a specific number to each farmer. This suggests increased control in the future, as individual farmers who create residue problems can be identified and disciplined individually.

Chayote Exporters' Forms of Control

There are a number of important structural differences between the minivegetable and chayote sectors that influence the contract relationship. First, the chayote sector depends almost entirely on one export variety, *quelite,* and its exporters focus only on chayote. This contrasts with the minivegetable exporters, who often handle dozens of different types of produce, although only a few crop types are exported. Second, while minivegetable exporters do not grow any produce, the larger chayote exporters are packer-shippers; they contract with farmers and also own chayote farms, farming substantial areas and producing about half of their exported chayote. Third, minivegetable exporters offer farmers a fixed and generally upward-trending price, which was about US$1 per kg of produce in 2003. Chayote exporters, in contrast, offer a variable price that reflects the price in the United States. This price has trended downward since the mid-1990s (figure 4.2).

During the interviews, I asked managers to explain past residue rejections. ChayoRico's manager began with the idea that chayote is a very minor crop to the pesticide industry, the yearly production of which pales in comparison to other commodities such as rice and pineapple in Costa Rica. He pointed out that malathion, an organophosphate insecticide, is the only EPA-registered pesticide specifically for chayote and that it is very unlikely that agrochemical companies will come to Costa Rica to conduct pesticide registration studies. He explained that the lack of registered products means that violations can happen easily:

We have a pest of [white]flies, so we go and look for the product to apply for these [white]flies in tomato. And if it is good, we apply the

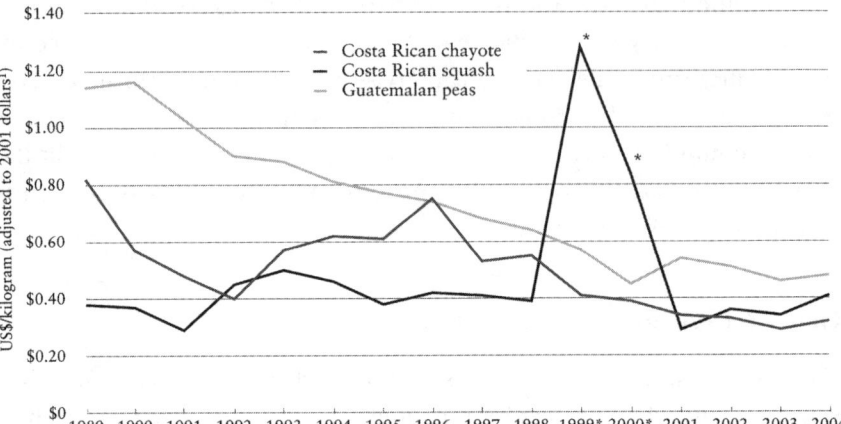

* The data for quantity of Costa Rican squash likely includes varieties in addition to the mini-squashes.
[1] Dollar conversions are from Inflation Calculator based on Robert Sahr's index at http://www.cjr.org/tools/inflation/.
Source: FAS 2005.

Figure 4.2   Prices of selected Central American vegetable exports at U.S. port of entry, 1989–2004.

pesticide for tomatoes [in chayote]. But if they do an analysis in the U.S. and this pesticide comes out, in the U.S. the person goes to the law and it says it is not authorized for this product. That is what is happening. And because of this we have detentions. OK. Now, there are strong products and weaker products in terms of residues. For example, with the weaker products, you have residuality of two or three days. You apply it, [and] in three days there are no residues. When the U.S. does an analysis, there is nothing there. But, for example, methamidophos has thirty-one days, I think it is, of residues. That's a lot of time, and it is very likely that your produce has residues [if farmers use methamidophos]. (ChayoRico manager, interview, November 2003)

Likewise, ChayotEx's manager mentioned the lack of registered agrochemicals as a problem for the sector: "On no labels, of any products, does it say it is for chayote. None come for chayote. None" (ChayotEx manager, interview, December 2003).

This is only partially true but reflects the lack of information available to exporters. During my field research in 2003 and 2004, I accessed the EPA's publicly available pesticide registration database. The only way of doing so involved an old and cumbersome Microsoft DOS (pre-Windows) program for which each user had to install the correct databases. When I performed a query for all agrochemicals registered on chayote by the EPA, the database indeed showed only malathion, which explains exporters' understandings of EPA tolerances. The program was flawed, however, since it did not show all of the pesticides registered for the cucurbit, squash, or summer squash groups, even though these would be permitted on chayote since the FDA considers chayote to be the same as summer squash when enforcing EPA tolerances (Carolyn Makovi, pers. comm., October 17, 2005).

The appearance that the EPA has only registered a single insecticide on chayote presents an important question for chayote exporters, since a single pesticide alone cannot control chayote's pest problems: which pesticides can be used if only one is registered? Exporters could tell farmers to stop pesticide use on chayote altogether, but given the prevalence of pesticide use in the area, this would probably be met with great skepticism and laughs. Instead, managers create their own guidelines. ChayoRico's manager uses crop similarities and PHIs as guidance for pesticide-use recommendations:

> We try to suggest to the farmer that they use products that they apply in other crops similar to chayote, like cucumber, like melon, like tomato, like ayote; the cucurbits.[8] OK, so there is nothing for chayote, but if these products are used in cucumber, perhaps they won't be so problematic in chayote. That is more or less the analogy that we are applying. Nothing that is more than three days . . . from the day of application to the harvest [PHI]. . . . [T]hey don't use products where the label says five or fifteen days. (ChayoRico manager, interview, November 2003)

ChayotEx compiled a list of insecticides that have been used in the area for years without causing rejections. If they are used according to the label, the manager of ChayotEx believes that they will not cause residue violations. The list is very small and consists of agrochemicals with low PHIs (cypermethrin, thiamethoxam, and avermectin), which should mean that they leave low or even nondetectable amounts of residues. This emphasis on less residual insecticides is the same as in the minivegetable sector.

Chayote exporters do not have written contracts with farmers. The relationship is more fluid than in the minivegetable sector, since chayote exporters do not require as much loyalty.[9] Some farmers sell to two exporters, and farmers can easily change exporters. The manager for ChayoRico described the verbal agreement in the following manner. "The farmer normally has produce [to sell] every week. Therefore, you have a regular relationship. . . . You don't have to renegotiate the price each week because you already know each other. It is a verbal contract, one of familiarity." I asked how the firm decides which farmers to purchase from. He replied that "if you were a new farmer that comes here and buys land and starts to plant, you will visit three or four packers and you tell them, 'I have chayote, do you want to buy from me?' and they will tell [you] their conditions. . . . But once we arrive on an agreement that you sell me the product, we know that we are partners" (ChayoRico manager, interview, November 2003). The manager of SuperChayote expressed a similar type of contractual relationship based on a verbal agreement. He also explained that the company purchases from a core group of farmers and was able to provide me a list of those farmers (SuperChayote manager, interview, December 2003). Similarly, ChayotEx's manager noted that there were no formal contracts but that they have verbal agreements with about twenty farmers (ChayotEx manager, interview, December 2003).

Chayote exporters are not as selective in choosing farmers as exporters in the minivegetable sector, and chayote farmers do not belong to farmer organizations that create within-class discipline. There is, however, some sectoral solidarity in which farmers understand that their use of problematic pesticides can decrease the international competitiveness of Costa Rican chayote. With these differences in mind, I asked managers of chayote exporters how they control pesticide use. The strategies include suggesting which pesticides to use, requiring that farmers fill out *registros*, visiting farms to ensure compliance, creating traceability systems, and conducting semirandom residue tests of selected farmers' produce. When asked about controlling pesticide use, ChayoRico's manager responded:

MANAGER, CHAYORICO: It is difficult, and we are almost sure that some farmers do . . . things that aren't included in good agricultural practices.

RG: Like I've heard that some farmers use methamidophos during harvest time. . . . What do you do to prevent this?

MANAGER, CHAYORICO: Random controls, random residue tests. . . . And as a farming business we know the prohibited practices.

RG: But it is difficult to control them?

MANAGER, CHAYORICO: It is very difficult, because culturally they believe that what they are doing is fine. Or they don't understand the reason for not using methamidophos.

RG: Do you talk to them about what they can and can't do? Is it a list of things, or only verbal [communication]?

MANAGER, CHAYORICO: No, no. There is a verbal part, the education of them, the communication [about what is permitted], and we give them certain *registros* that we want them to fill out. (ChayoRico manager, interview, November 2003)

I suspected that the residue tests were not truly random, so I aired my doubts. He responded:

MANAGER, CHAYORICO: No, those that we have suspicions about are chosen. Or from whatever farmer, from a bag we choose [the names of] five farmers, and we put the produce of the five mixed together and send it [to be tested for residues in San José]. If it turns out well, those five are OK. If it comes out badly, one of the five is bad.

RG: Has this happened in the past . . . [and] you had to cut off a relationship with some farmers?

MANAGER, CHAYORICO: Six years ago we had this problem, one was cut off. . . . After that, I believe we had one, if I remember, that MAG did an analysis from a [farmer's chayote] plant, it came out with problems, and we had to end the relationship. (ChayoRico manager, interview, November 2003)

In contrast, SuperChayote does not conduct residue tests of chayote.[10] I asked SuperChayote's manager how the company ensures that the farmers it buys from use GAP. He responded:

MANAGER, SUPERCHAYOTE: We are always monitoring their farms; we are
giving them chats about what [pesticides] they have to use, the good
agricultural practices that they have to use in the field. And since it is
a small area, we have them very controlled. . . . But . . . I don't think
that they cause many problems with chayote. Since everyone is so
close, we have them very controlled.

RG: What is the monitoring that you do?

MANAGER, SUPERCHAYOTE: Well, control of chemical products that they
are using. That the water isn't contaminated at the time of spraying, that the workers have their latrines in the field, that they wash
their hands when they harvest the product. The waste [containers]
from the chemical products, that they collect them and store them
and return to the vendors so that they don't have problems with the
environment, right?

RG: Is this monitoring every week, or every month, or what?

MANAGER, SUPERCHAYOTE: We do it every two weeks. . . . As I've said,
the producer is really very conscientious about this. (SuperChayote
manager, interview, December 2003)

ChayotEx's manager, when asked how she ensures that farmers use GAP,
mentioned the same strategies as SuperChayote's manager: *registros,*
farm visits, and talks for farmers. In the *registros,* ChayotEx requires
farmers to take meticulous notes: "Everything they do, the normal [prac-
tices], like irrigating, etc., and the days they harvest chayote, and, princi-
pally, the use of agrochemicals. They have to put the day, such and such
product, up to the hour, the date, the quantity, etc. And [the farm inspec-
tor] uses an hour to visit them to implement the control. Yes, we have
control of what they are doing, collecting these pages [of the *registro*], so
we can archive them, and so that they don't use an agrochemical that is
not permitted." The control, she mentioned, is to enforce their conscien-
tiousness. "They know to not use [prohibited products] because we are
not going to buy their chayote. They have to be responsible" (ChayotEx
manager, interview, December 2003). ChayotEx was also starting a residue analysis plan, coordinated with MAG, that will happen on a weekly
basis.

All three of the chayote exporters I interviewed had implemented a system of traceability. ChayoRico began a traceability system in May 2003, along with a stronger emphasis on GAP. During the time of my interviews, December 2003, SuperChayote and ChayotEx had both recently started a traceability system.

While much of the discussion during the interviews gave the impression that all was well in the chayote sector, I probed about problems in pesticide policing, especially regarding methamidophos. The use of methamidophos and other residual organophosphate insecticides by chayote farmers arose in conversations with farmers and exporters, as methamidophos has a long history of causing violations in chayote exports starting in the mid-1980s (see chapter 3).[11] I asked the manager of ChayoRico why some farmers continue to use very residual pesticides:

MANAGER, CHAYORICO: Because they believe . . . it has better results.

RG: I've heard that . . . with methamidophos one can harvest more.

MANAGER, CHAYORICO: I don't believe that. Methamidophos is an insecticide; it has nothing to do with chemical properties. But yes, it gives that produce a better quality—smooth, without insect damage—at a very low cost. Methamidophos is very cheap. (ChayoRico manager, interview, November 2003)

The employee who inspects the farms of ChayotEx's chayote growers mentioned the same reason while he was in the room during my interview with the firms' manager. The exchange illustrates the contradictions inherent in trying to persuade farmers to use less residual pesticides in an export sector in which low chayote prices have become the norm (see figure 4.2).

FARM INSPECTOR, CHAYOTEX: Methamidophos is cheap, and effective. . . . It kills whatever pest you have in your chayote field; . . . since it works so well, you have a good harvest. The production . . . is almost double the chayote that you can harvest normally [without methamidophos].

MANAGER, CHAYOTEX: But now it is prohibited. In chayote it is prohibited because the residuality in chayote is a month. . . .

FARM INSPECTOR, CHAYOTEX: Methamidophos lasts a month . . . and twice per week you harvest chayote. With tomato, celery, peppers, at the beginning when the plant is small, everyone uses methamidophos. In chayote the problem is that there are farmers who are not responsible. And they use it, and they don't say anything to you.

MANAGER, CHAYOTEX: Yes, because, for example, this list [of permitted pesticides] that I showed you, these [approved] insecticides are very, very expensive. So, we require of them "Look, use thiamethoxam, use cypermethrin." They say, "You're crazy, they are extremely expensive and chayote is cheap [the farm-gate price is low]." It is very difficult, because on the one hand we require them to use this [pesticide] that is very expensive, and on the other hand they are going to say to me, "Pay me well," and on the other hand they [produce buyers in the United States] are paying us poorly. We are going through a crisis here in chayote because of this, because it is cheap, because the inputs are very expensive, and the farmer is asking us to understand. It is unfortunate. . . . But, lamentably, now you have to tell them, "If you use this [methamidophos], I will not buy chayote from you, and nobody else will buy it either." (ChayotEx manager, interview, December 2003)

The manager went on to explain that she belongs to a chayote exporters' council that shares information on farmers whose produce has had illegal residues, so this is not an idle threat. The export price is consistently higher than the national market price even though the export price fluctuates, so losing access to the export market can be a serious financial blow to export chayote growers.

Many of the types of control used by chayote exporters had only recently been adopted in response to the requirements of the U.S. Bioterrorism Act. Control by exporters was more informal previously, while MAG was more responsible for formal policing. MAG emphasized that farmers must avoid methamidophos use in chayote, since this was causing residue violations in the United States. The state's involvement speaks volumes about the importance of export crops to foreign exchange earnings and debt service as part of structural adjustment (see Conroy, Murray, and Rosset 1996). Chayote exports to the United States in 2002 were 22,080 metric tons, worth $7.4 million (United States Department of Agriculture

Foreign Agricultural Service 2005). The chayote sector is larger than the minivegetable sector with more farmers and exporters involved, making the state's engagement in protecting capital accumulation more likely. According to the manager of ChayotEx,

> Before, it was verbal more than anything. At the level of Ujarrás [Valley], "Don't use methamidophos, don't use this, don't use this other one," products that everyone knows [not to use]. However, what the others used could put you in danger, and they used them, and we didn't know. But, for example, here at the level of the business we held a conference [about] which are the products to use; we explain. But right now. Not before. Before it was controlled through MAG because they had a program of chemical residue analysis. And you would see them every week taking samples from all of Ujarrás [Valley] . . . because there was much use of methamidophos, a very bad product. And then, once they controlled the situation, and they figured that the majority came out free [of residues], they said that the people understood and that they were working well. But before. It was a very strict program that already ended. They still check; once in a while they come and take a sample of five farmers. (ChayotEx manager, interview, December 2003)

Exporters and farmers discussed MAG's participation as a strict monitoring of methamidophos use around the Ujarrás Valley. When I asked chayote farmers if they had had produce sampled for residues, many responded that personnel from MAG had shown up one day and taken samples. Almost no farmers heard back from MAG, a positive outcome since the policy was to only return to quarantine the field if methamidophos residues were found. One farmer in Río Regado explained how this happened to a farmer down the street from him (Farmer survey, July 2003). The incident appeared on television, likely striking fear into other chayote farmers that they would be caught using methamidophos, be publicly shamed, and lose access to the export market.

The strong but simple message about pesticides from MAG and chayote exporters is to avoid organophosphate insecticides that caused previous violations, especially methamidophos. As is partly the case in minivegetables, exporters and MAG are much less concerned about fungicides, since these have not yet caused residue violations for local produce.

Pesticide Residues as Threats to Capital Accumulation

Pesticide residue violation rates in Costa Rican vegetable exports are below the rates of many other fresh vegetable export sectors in Central America and the rest of the world (see table 4.1) and are far from the crisis levels that Guatemala and the Dominican Republic faced in the 1980s and early 1990s (Conroy, Murray, and Rosset 1996). This is rather remarkable given that Costa Rican agriculture is the most pesticide intensive on the planet. Part of the reason for this relative success is exporters' attempts to control export farmers' pesticide use.

By focusing on exporters' direct fashioning of the production process through contract farming, this chapter highlights the main reasons behind the rationalization[12] of pesticide use by export farmers. Exporters as off-farm capital are economically compelled to understand the regulatory standards and systems in their commodity chains. In the context of a paucity of information, exporter employees use what information is available—pesticide labels, crop similarities, incomplete information from the EPA on registered pesticides, and past rejections—to make judgments about which pesticides they should prohibit and which ones they should prescribe to their contract farmers. In both the chayote and mini-vegetable sectors, they create varying regimes of social control—combining verbal agreements, written contracts, the posting of GAP lists, visits by agronomists and farm inspectors for prescriptions and enforcement, the establishment of farmer associations to create peer pressure, intermittent residue testing, and sanctions for farmers who violate their standards—to directly shape production on the farm so that vegetables comply with regulations from afar. In the chayote sector, the Costa Rican state has played an important role through its residue testing program to create sanctions for the use of methamidophos on export chayote. In these ways, the exporter-farmer relationship is much like other attempts at workforce standardization such as Taylorism, which "defined certain behavior as moral and other behaviors as immoral. . . . [B]oth negative and positive sanctions were used to enforce these standards. Model workers would be rewarded not only economically but also by the prestige attached to recognition. In contrast, those who violated the standards would be punished, in extreme cases by banishment from the company.

This is not to suggest that resistance was never encountered. Indeed, in every location it was instituted, such attempts at standardization were *projects,* not faits accomplis" (Busch 2011, 124, emphasis in original).

While these forms of control are important, they are certainly not a done deal, as Busch (2011) notes. The message from exporters and the state about pesticides is simplified—in relation to actual regulations—and historically contingent. Exporters and relevant government entities in Costa Rica do not completely understand U.S. pesticide residue regulations and do not receive communication from the EPA or the FDA to tell them which pesticides can be used on which crops or to tell them about specific instances of illegal residues being detected. Exporters must depend on importers' word-of-mouth communication or sometimes faulty stories in the news media for information on pesticide tolerances and violation incidents. Export firm employees must then use this incomplete knowledge of pesticide residue requirements to inform their standards. This is one reason why they simplify—and sometimes deliberately misrepresent—the requirements when they communicate with farmers for the purpose of minimizing pesticide residue violations. Export firm employees then attempt to police these greatly simplified standards about pesticide residue regulations. Past residue problems have a strong influence on what exporters decide to prohibit and prescribe, demonstrating that market requirements are socially mediated in sector- and location-specific ways.

Previous studies of pesticide use on agroexports have not empirically examined exporters' understandings and actions or the relationship between exporters and contract farmers. Instead, they have assumed that exporters demand perfect-looking produce but pay little attention to pesticide residues (Wright 1990) and have also assumed that if the agroexport sector experiences pesticide problems, exporters can simply pack up and start business elsewhere (Murray and Hoppin 1992). With these assumptions, they missed the possibility of regulatory adaptation, that is, that exporters may attempt to learn from residue violations and make changes in the social arrangements governing the production system. Importantly, the export firms included in the study are all owned by Costa Ricans with substantial ties to their communities. This means that they are relatively fixed in place and have a stake in the future success of the agroexport sectors in their present location. These factors help explain why these exporters did not simply leave and take their export crops into new regions without a history of high pesticide use, as in other Latin American agroexport sectors and as we might expect from transnational corporations. Far from ignoring the residue problem, various actors in export commodity chains in Costa Rica—from exporter management to small-scale farmers—have attempted to adapt to the North's regulatory demands because of the potential for large financial losses.

Residue violations on exports have also prompted state action, including outright pesticide bans such as with DDT and aldicarb in Costa Rica, while the pesticide poisoning and chronic diseases of workers, farmers, and even children is basically irrelevant to the state's regulatory logic. These findings parallel Blaikie's (1985) argument about the political ecology of soils: only when soil erosion affects the prospects of capital accumulation will efforts be made to try to solve it. These export sectors show clearly that when capital accumulation and export earnings are threatened by pesticide residue violations, firms and the state take pesticide use seriously but only as it relates to residues.

Contract farmers who continue producing the new agroexports must walk the tricky line between producing vegetables of high enough quality for the export market and not causing rejections due to illegal residues, all while trying to make a living—keeping their costs of production lower than their return from vegetable sales. If exporters squeeze too hard by not offering high enough prices, farmers' situations can become economically untenable, and conflicts can manifest, overtly and covertly.

Off-farm capital and farmers have very different responses in these situations. One strategy that exporters use to enforce pesticide rationalization is summarily excluding smallholders from their contracts. Facing stronger food standards, exporters in many African countries that provision European markets have increased reliance on large-scale farming operations to the detriment of smallholders (Barrett et al. 1999; Dolan and Humphrey 2000; Freidberg 2003, 2004). This means greater social exclusion in new agroexport production and likely increased inequality, since excluding small farmers takes away their most lucrative market.

The Costa Rican case shows that the inclusion of smallholders in new agroexport production is largely in the hands of the management personnel of agroexport firms. The Costa Rican minivegetable sector remains inclusive of smallholders because of the strong commitment of one exporting firm. At the helm of the firm is a so-called benign dictator, as also identified in other work on new agroexports (Freidberg 2004), who remains interested in new agroexport production as a means of increasing incomes of smallholders in the face of difficulties in the national market, especially its fluctuating prices. The chayote export sector also still involves smallholders because the export firms' farming operations do not produce enough to satisfy market demand and because chayote and its seeds, as a nonhybridized crop, remain in farmers' control. There is, however, no guarantee that contracts with these smaller farmers will be maintained if the exporters can meet the demand through their own farming operations. Overall, then, agroexport production in Northern Cartago and the Ujarrás Valley has been fairly inclusive of small-scale farmers and has not created widespread social disruption, as did the Central American cattle boom (Williams 1986).

This inclusion of small-scale contract farmers, however, is not a necessary condition of new agroexport production and could well shift. The Bioterrorism Act of 2003 may be inducing changes in agroexport sectors around the world, and these rearrangements may have negative implications for export farmers. In the study site, exporters' control became stronger with tighter regulation of the U.S. food supply. The implementation of traceability systems—as started by Costa Rican exporters shortly after the passage of the Bioterrorism Act—has potentially large ramifications. Before traceability systems, farmers could rely on some anonymity since their produce was packed with everyone else's. Traceability may allow exporters to pass the costs of losses on to individual farmers, as has already happened in some export sectors in Latin America, such as with Chilean table grapes (Murray 1997). Since studies have not yet examined the social inclusion and exclusion prompted by these changes, important questions remain about the many consequences of export firms' strategies of control in relation to tightening food standards.

On the other hand, contract farmers can respond to unjust pesticide policing, which is evident when efforts by off-farm capital to rationalize pesticide use fail. The case of vegetable exports from Guatemala—with a consistently high violation rate of 18 percent for all vegetables and 29 percent for peas from 1996 to 2006 (Galt 2010)—shows that stricter control alone will not create the outcome desired by exporters. Farmers there have continued to use prohibited pesticides even after substantial efforts to change their pesticide use (Julian, Sullivan, and Sánchez 2000; Thrupp, Bergeron, and Waters 1995). Many Guatemalan agroexport sectors rely mostly on spot market provisions (rather than coordinated supply chains), thereby preventing the kind of direct fashioning of the production process through contract farming that we see in Costa Rica, and that is increasingly necessary for export markets (Galt 2010).

Additionally, using illegal pesticides seems a likely farmer response to declining farm-gate prices, one that is both economically rational and might be a means of resistance to exploitative social relations. Contract farmers can employ "weapons of the weak" (Scott 1985) by using pesticides that cause rejections. Peasant resistance to coercive state conservation programs, especially resistance manifested in the destruction of protected forests by arson around the world, has been well documented (Agrawal 2005; Kuhlken 1999). Using highly residual pesticides as a form of resistance is similar to these acts of arson: it damages the shared resource (export market/forest), which harms those who are seen to directly benefit most from it (exporters/the state or colonial power) but also those who depend on the resource for their livelihoods (export farmers/ forest resource users). Farmers may use prohibited pesticides as a form of resistance because of their loss of decision-making autonomy or because of animosity over what they see as the exporter's exploitative pricing policies. Consequently, violation rates may paradoxically go up when pesticide policing is increased if farmers see these as cutting into their well-being.

Thus, off-farm capital and the state do not and cannot solely determine farmers' pesticide use. These off-farm actors set the parameters around which pesticides farmers can use and how, but what happens in farm fields is largely outside of their control. This supports Grossman's (1998, 189) conclusion that "although capital and the state attempt to instill particular, uniform regimes of agrochemical use, what actually happens at the local level is much more complex." In part this is because of ever partial control over land users by off-farm capital and the state in contract farming (Clapp 1994), meaning that farmers have considerable opportunities to go against expected or required practices.

Whether farmers decide to go along with pesticide policing is difficult to predict because it depends on how a number of tensions are resolved. One of these tensions is whether farmers are well compensated for their produce, which is certainly not a guarantee in agroexport production. Another is how export farmers navigate the tension of producing vegetables for aesthetically demanding markets while also minimizing pesticide residues. The next chapter examines this last tension and compares farmers' performance to actual regulations.

5 Regulatory Risk and the Temptations of Methamidophos

For one of my first farmer surveys, I was searching for José Méndez, a medium-scale farmer who produces minisquash and green beans for export. I found him working in one of his squash fields. A very cheerful guy, José seemed genuinely happy to meet with me. As I called out to him and walked toward him, he yelled back, with a big smile, "Tell me what you need. You guys command, and we deliver." By "you guys" he meant the United States generally and its market requirements, which for him are mediated by the exporter to which he sells. From our conversation, he clearly viewed growing for the U.S. market as very demanding, requiring caution with pesticides and carefully selecting, handling, and washing produce. The exporter's forms of control were clearly at the forefront of his mind as we spoke about farming for the export market. As my fieldwork progressed, I found this a common view among contract farmers growing for export.

The story of meeting José highlights the ways in which food regulations in the industrialized world increasingly affect food-producing industries and farmers in developing countries. As noted in the Introduction, researchers in the 1990s and early 2000s typically found the transformations from new agroexports to be socially and environmentally negative in both Africa (Barrett et al. 1999; Freidberg 2003; Opondo 2000) and Latin America (Murray and Hoppin 1992). However, the development of alternative agrifood networks—including organic, fair trade, and eco-certification—may have positive effects on farmers' production practices and communities in the global South (e.g., Jaffee 2007; Melo and Wolf 2005). Additionally, important segments of conventional agrifood systems have undergone recent change and demand better environmental and social performance, such as the Global Partnership for Safe and Sustainable Agriculture (originally Euro-Retailer Produce Working Group

Good Agricultural Practices and commonly known as EUREPGAP) standards created by European supermarket chains, private-interest regulation by supermarkets in the United Kingdom, and vertically integrated agroexport firms in the Africa-to-Europe fresh produce trade (Barrett et al. 1999; Flynn, Marsden, and Whatmore 1994; Freidberg 2003; Konefal, Mascarenhas, and Hatanaka 2005; Marsden, Flynn, and Harrison 1999; Winter 1997).

In this age of certification and private standard proliferation in the South-to-North agroexport trade, I find it important to emphasize that nation-states are still responsible for much agrifood regulation in the global North. Even when private standards become dominant, state standards are still a baseline to which produce must correspond. In acting from afar through the commodity chain as noted in chapter 4, state-level standards, often together with private standards, apply to almost all produce imported from the global South.[1] The enforcement of these standards creates what I call regulatory risk—the possibility that an actor's behavior will be subject to regulation and that out-of-compliance behavior will result in negative consequences that impact the actor. In short, regulatory risk is the risk of being sanctioned either by the state or by other actors in the commodity chain that act as enforcers. Despite its global reach, from Argentinean apples to Zimbabwean zucchini, we know little about how producers and contractors experience regulatory risk in relation to their production practices. We know less about the environmental consequences of these regulations in places in the global South where enforcement touches down. A thick description and theorization of these effects is thus necessary to understand the functioning and contradictions of transnational food system governance of conventional production.

Discussions of agrifood regulation remain mostly disconnected from the literature on pesticide use in developing countries. As Grossman (1998) points out, the vast majority of this literature emphasizes that farmers indiscriminately use and abuse dangerous pesticides and exercise almost no caution (Abeysekera 1988; Feola and Binder 2010; Guan-Soon and Seng-Hock 1987; Heong, Escalada, and Lazaro 1995; Hui et al. 2003; Yen, Bekele, and Kalloo 1999; Zaidi 1984). Studies point to the lack of protective clothing worn, the common use of overdoses (spraying more than recommended) and pesticide cocktails, the lack of respect for the time required between application and harvest (the preharvest interval [PHI]), and the imprudent use of highly toxic pesticides such as the organophosphate and carbamate insecticides.

This chapter, then, connects agrifood regulation with farmer pesticide use. Having examined the socioecological dynamics of the study site and how pesticide regulations are translated by exporters, I now turn to farmers' land-use decisions as they relate to pest control. The questions addressed are: What are the effects of pesticide residue regulations, as mediated by exporters, on farmers' production practices? And, in light of arguments around contract farmers being unfree in their decision-making processes (Watts 1992), to what extent are contract farmers producing export crops in compliance with export market regulations?

Theorizing Regulatory Risk

Risks from technologies such as pesticides feature prominently in the rise of the environmental movement (Carson 1994) and in the risk society thesis (Beck 1992), where environmental risks created by modern industrial society increasingly dominate social debate. These include human health risks to workers and farmers in direct contact with pesticides, such as poisoning, neurological damage, cancer, endocrine system disruption, and immune system suppression, and to consumers who can encounter agrochemicals in their drinking water, food, and environment. Environmental risks are clearly important as well (Pimentel and Lehman 1993).

In addition to literature addressing risk from pesticides as agricultural externalities, there is a vast literature from many disciplines that addresses farmers' and rural communities' responses to risks in agriculture. For example, researchers have shown that communities have developed social structures to act as a resource reservoir for those households experiencing crop failure (Kirkby 1974) and for more widespread events such as severe droughts (Watts 1983b). Political ecologists have examined farmers' reasons behind planting diverse crops, showing that one rationale is hedging against crop failure in risky mountain climates (Zimmerer 1996). In response to agricultural marketing risks, small farmers plant a combination of high-risk cash crops and lower-risk subsistence crops

(Barham, Carter, and Sigelko 1995; Feder 1980) and maintain social relationships in different marketing channels (Mannon 2005).

In addition to these understandings of risk in agriculture, we need to consider farmers' responses to a relatively new and underappreciated form of risk with a very different geographic dimension: *regulatory risk*. As noted above, regulatory risk arises from the enforcement of extraeconomic market requirements in agricultural commodity markets and is most notable in the form of phytosanitary and food safety regulation in the industrialized nations.[2] In addition, regulatory risk is borne by actors in commodity chains and acts on them from afar.

Within the export commodity chains in the study site, regulatory risk is borne at two levels—by exporters and by farmers—and the stakes for each are quite high. As seen in chapter 4, exporters face the loss of a shipment if it is found to violate regulatory standards in the imported country. This is a direct economic loss and can also destroy their relationship with importers. Thus, exporters establish pesticide policing mechanisms, which are the regimes that instill regulatory risk for contract farmers.

For contract farmers, regulatory risk adds to the already substantial number of risks they face. As noted in chapter 2, contract farming allows for family farms to become functional to off-farm capital, in large part because it ensures that farmers continue to bear many risks of farming, including pest outbreaks, drought, frost, and floods. Regulatory risk adds to these risks by potentially resulting in the farmer's loss of access to more lucrative export market channels. This expression of regulatory risk means that farmers can face a direct economic loss if an exporter finds them at fault for regulatory violations. In other commodity chains, farmers bear even more risks, such as the "external catastrophe" clause in some Chilean grape export contracts. This clause hands ownership of the shipment back to farmers "in the case of sudden embargoes due to war, hijack or any other factor" (Murray 1997, 49).

Regulatory risk that farmers face is an on-the-ground consequence of the changing spatiality of agrifood governance. Whereas land users have been historically sanctioned by communities governing the commons (Blaikie and Brookfield 1987; Ostrom and Gardner 1993), regulatory risk arising from regulation from afar and mediated through social and economic relationships is a recent form of sanction in agriculture resulting from the new agroexport booms around the world that provision an increasingly vigilant global North.

Previous research on pesticide use vis-à-vis regulation from afar suffers from weak theorization concerning export farmers' decision making in the context of regulatory risk. Most literature on Latin American new agroexports implies that export farmers do not take these regulatory risks seriously because they are not informed about them or must ignore them because they find themselves on the pesticide treadmill or struggling to meet cosmetic requirements (Conroy, Murray, and Rosset 1996; Hamilton and Fischer 2003; Murray 1994; Thrupp, Bergeron, and Waters 1995). In his case study of export vegetable production in Mexico, Wright (1986, 51) points out that "although only 1 of 15 loads of vegetables from Culiacan is tested by the FDA [U.S. Food and Drug Administration] for pesticide residues, every single load is examined for conformity to cosmetic standards. Thus, the incentive given by the US market for higher pesticide use is more compelling than the incentives for lower pesticide use created by pesticide residue standards." Similarly, Thrupp, Bergeron, and Waters (1995, 51) state that "the immediate pressures to increase pesticide use tend to outweigh other considerations." We do not know why this would be the case, how it may be prevented, or how it is impacted by various human and nonhuman actors in the commodity chain. My findings from chapter 4 suggest that outcomes will be shaped by a crucial but neglected element: the social relations of exchange in contract farming. In many commodity chains, exporters impose regulatory risk through policing of land-use practices and attempting to directly shape the production process (Watts 1992; Wolf, Hueth, and Ligon 2001), with considerable attention at times paid to farmers' pesticide use.

Although exporters in the study site exert various forms of control over contract farmers' pesticide use, policing is not ever-present—that is, the export firms' agronomists only visits periodically (if at all), and Ministry of Agriculture and Livestock (MAG) personnel only visit sporadically and sample from a handful of chayote farms. Farmers therefore still have the option of using proscribed pesticides that remain available for sale. Contract farmers can decide to use pesticides that they know violate regulations, thereby gaining potential productivity benefits, but they run the risk of being caught and sanctioned. While causing the loss of a shipment can lead to substantial monetary losses to exporters, these losses might be passed on to the exporters' farmers through direct price cuts or through a loss in market share for Costa Rican produce that would decrease the overall volume that local export farmers could ship. Additionally, individual farmers found to be responsible for a violation will lose access to the higher-paying export market.

As exporters noted, despite the control they exert over contract farmers, there are still economic reasons for farmers to use prohibited pesticides such as methamidophos. Methamidophos has a reputation of being an extremely powerful insecticide, and farmers say that it can double yields of high-quality chayote. Thus, even in export systems where farmers face sanction as a result of using this and related pesticides, there are short-term economic reasons to use some proscribed pesticides. Pesticide-use decisions therefore must be understood in the context of contradictory economic pressures—using prohibited pesticides can result in short-term gain but with the risk of serious long-term loss—*and* contradictory regulatory pressures—for cosmetically perfect produce free of pests and diseases but without (illegal) residues—that are shaped by broader social relations of capital accumulation and the governance of trade and economic relations. How these play out in farmers' fields is not easy to predict.

Farmers' Caution Concerning Residues

In the face of regulatory risk, farmers can choose to exercise caution in their use of pesticides. While most researchers treat caution with agrochemicals as if it is a binary variable—that is, farmers are cautious or not cautious—we must disaggregate it to better understand it. Since pesticides cause a multitude of problems—including compromised health for farmers, workers, rural residents, and consumers in addition to environmental problems including wildlife and fish kills, surface water contamination, and groundwater contamination—there are many different types of caution that can help ameliorate these different problems. These include caution about (1) immediate exposure to those people in the field during the application, (2) exposure of those living nearby, (3) pesticide residues eaten by those who consume the produce, and (4) the release of pesticides into the environment, which itself can be divided into subcategories, including protecting groundwater, surface water, beneficial insects, and wildlife. For each of the many problems caused by pesticides, one could list precautionary measures that could decrease or avoid it. Thus, with some effort researchers will find almost all farmers who use pesticides to be exercising some form of caution while ignoring other types of caution.

My focus is on farmers' caution concerning pesticide residues on their harvested produce as a response to regulation from afar. Many factors affect the amount of a pesticide that remains as residue on food. These include a farmer's decisions about which pesticide to apply, how much to apply, the frequency of applications of the same pesticide, the time between spraying and harvesting (the PHI), and various other factors generally not under the farmer's control, including weather conditions during and after the application, the characteristics of the chemical, and environmental and crop characteristics that determine rate of breakdown (Wargo 1998). This chapter focuses on those factors directly under the farmer's control: the specific pesticides applied, the dose (application rate) used, and the PHI. I use data from my farmer surveys to compare export farmers' pesticide use in the study site to two reference points. The first reference point is the typical farmer in developing countries as portrayed by the literature. Grossman's (1992, 1998) political ecological work on pesticides implicitly uses the same reference point. These typical farmers often depend on highly toxic organophosphate and carbamate insecticides (e.g., Wright 1986, 1990), use overdoses, and disregard PHIs (Abeysekera 1988; Yen, Bekele, and Kalloo 1999). The second reference point I use for comparison is U.S. regulation, specifically U.S. Environmental Protection Agency (EPA) pesticide tolerances (Environmental Protection Agency 2004) and FDA enforcement for the crops in question (Food and Drug Administration 2013). This reference point allows for an understanding of regulatory compliance and failures.

Most export farmers expressed concern about the possibility of causing rejections due to pesticide residues and noted that producing for export has made them more aware of pesticide residues. The level of concern has increased over time. In the 1970s, the early days of chayote exports, it is unlikely that farmers were seriously concerned with pesticide residues on their produce since it was a relatively unknown risk, the negative consequences of which had not been felt. The rejections of chayote in the

1980s caused substantial financial losses for exporters and created the impetus to police pesticide use, as detailed in chapter 4. With increased monitoring of pesticide use by export firms and the state, export farmers had to start weighing regulatory risks in their decisions starting in the 1980s. Additionally, the minivegetable export sector's early emphasis on zero residues made its contract farmers aware from the beginning.

Many export farmers explained that the switch to export production required being more cautious with pesticide use. For example, Rodrigo, who used to grow vegetables for the national market, described the changes when he started growing minisquash for export:

RODRIGO: [It was] in chemicals, more than anything. Before I used more residual products. . . .

RG: Why did you change?

RODRIGO: Because it was a requirement that Mini-Horta demanded of one. Because it is a product for export. It is for export, so you have to lower the level of residuality of the chemical products. And also for one's own benefit.

RG: If you had not converted to nontraditional [export] products, how would your agricultural production practices be now?

RODRIGO: I would have continued using highly residual [pesticides]. Because it is one way that we can exist with the prices of the commodities and the pests that there are. The prices are low [for national market vegetables], and the pests appear every day. So, you have to be spraying, spraying products that are residual because you cannot get by [without them because they are more effective than less residual ones]. (In-depth interview, December 2003)

This was a common story in my follow-up interviews with export farmers. Similarly, Breslin (1996, 32) quotes minisquash farmer Martín Aguilar of Cipreses, Costa Rica, as saying, "Exporting is what puts the quality in our products. . . . Exporting gives us standards we have to meet."

This was confirmed through my interviews with pesticide salespeople. Among other questions, I asked if they noticed whether any farmers were particularly concerned about the residues or the PHI of agrochemicals they were spraying. In one instance, Saúl, a tall and lanky guy, responded, "Yes, the export farmers. They ask a lot about the residues that pesticides leave." The other pesticide salespeople I interviewed also noted that export farmers asked more questions about the residues that pesticides left on crops. These conversations serve as triangulation, which involves collecting data in different ways and from people in different social positions. These data are important checks on how exporters and farmers growing for different markets were portraying their caution with pesticides to me.

Yet choosing pesticides is not merely a matter of farmers trying their best to comply with regulation from afar. While many emphasized that they do their best, a few noted the temptation to use methamidophos, a highly toxic and effective insecticide not registered for use on squash in the United States or Canada. Many farmers and exporters noted that using methamidophos results in larger harvests, yet it is risky—vis-à-vis regulation—since it has caused past violations. Thus, spraying methamidophos, and organophosphates generally, involves risk calculation of a low-probability but high-consequence event for export farmers in the study site: is increased production worth the possibility of losing access to the export market? Importantly, the perception of the risk is amplified through its communication from exporters to farmers. Specifically, minivegetable exporters simplify the message about residues, saying that they do not allow *any* residues. Chayote exporters, on the other hand, target methamidophos, dimethoate, and the organophosphates generally. How do exporters' messages and forms of control affect the specific ways in which contract farmers use pesticides?

To answer this question, I rely on data about the pesticide use of fifteen farmers who grow minisquash for exporters in Cipreses and twenty chayote farmers who sell to the exporters in the Ujarrás Valley.[3] These data are summarized in table 5.1 and in figure A.1 in appendix 2. Table 5.1 reveals that organophosphate and carbamate insecticides, which are generally highly toxic and tend to dominate insecticide use in developing countries, account for 14.9 percent of all insecticide doses used during the growing cycle for export minisquash farmers. The generally less toxic and less residual pyrethroids are the dominant insecticides, making up 39.3 percent of all insecticide doses. For chayote farmers, pyrethroids make up two-thirds of all insecticide applications, while organophosphates and carbamates make up only 17.6 percent of the total.

Table 5.1 Insecticide classes used on export minisquash and chayote, Northern Cartago and the Ujarrás Valley, 2003–2004

	Export Minisquash (n=28)[a]	Export Chayote (n=20)
Total insecticide doses per crop cycle (mean)	9.7	37.7
Organophosphate and carbamates:		
Number of doses (mean)	1.7	8
Doses as a percentage of insecticide doses	14.9%	17.6%
Pyrethroids:		
Number of doses (mean)	3.8	26.2
Doses as a percentage of insecticide doses	39.3%	66.7%

[a] Some farmers contribute two to the sample number if they grow both scallop squash and zucchini.
Source: Author's farmer surveys, 2003–2004.

Export farmers generally follow pesticide label recommendations, in contrast to descriptions in the literature. Export squash farmers generally comply with the PHI, while export chayote farmers' compliance is lower. But while export minisquash farmers generally comply with insecticide PHIs, thereby demonstrating considerable caution in terms of avoiding high levels of insecticide residues on their produce, they generally do not comply with fungicide PHIs. The pattern is similar for export chayote farmers, but compliance is generally lower, especially for fungicides, with only about 5.6 percent of fungicides' PHIs being respected.

Export farmers also generally comply with dose recommendations. Of the fifty-nine pesticide active ingredients used on export minisquash, the average dose is within the recommended dose for all but three. On average, export chayote farmers use overdoses of four of forty-seven pesticide active ingredients. Most farmers use doses below the maximum dose recommendation. Even though farmers might be following the label, this does not mean that they are complying with regulation from afar, since Costa Rican and U.S. laws are substantially different.

Overall, the correspondence with U.S. regulation is not stellar. Only about half of the pesticides used conform to or are exempt from EPA tolerances on the specific crops in question. For both crops, the vast majority of pesticides are registered for agricultural use by the EPA on some crops in the United States but not for the squashes in question (U.S. pesticide registration occurs in specific crop-pesticide combinations). Farmers are also using pesticides not registered by the EPA, but these are ones that have never been registered[4] rather than pesticides that have been banned. Interestingly, the 2003 FDA tests for residues on these crops would have found only a small percentage of the pesticides actually used: 20.3 percent in minisquash and 31.9 percent in chayote (for more detail, see Galt 2009). Overall, while minisquash and chayote farmers demonstrate caution about pesticide residues (as suggested by the data on PHIs and dose), their pesticide use does not correspond very closely to U.S. regulation, and the majority of pesticide residues would be missed by FDA testing.

Although the general tendency is for farmers to try to comply with their exporters' requirements, some farmers use pesticides that they know can cause regulatory violations but do so in a way to minimize this risk. Phorate, a granulated organophosphate, is one such pesticide. It is used to keep *joboto*, the Costa Rican name for *Phyllophaga* spp. larvae, from eating crop roots. Farmers know the risk of causing violations but feel that they have no viable alternative. Their solution to this dilemma is to use very low doses when they violate the PHI, reasoning that a substantially lower dose that violates the PHI will leave low enough residues to escape detection (see Appendix 2, figure A.1). While logical, this is risky since any amount of phorate detected by the FDA will cause a violation. When it comes to methamidophos, the pesticide historically responsible for the most violations, some farmers still use it despite exporters' prohibitions against it. However, farmers are cautious about residues in terms of both PHI (using it only near planting time and not when the plant is producing fruit) and dose, using below what is recommended on the label.

Disconnections among Understandings, Practices, and Regulations

The way in which export farmers use pesticides in the study site is very different from how the majority of the literature describes farmers' pesticide use in developing countries. First, while they use some highly toxic organophosphate and carbamate insecticides, these do not dominate.

Instead, export farmers rely much more on pyrethroids, which have much shorter PHIs on average because they degrade faster. Second, most export farmers follow the recommended dose for all types of pesticides, and many export farmers intentionally use low doses in order to minimize residues.[5] Finally, both export minisquash and chayote farmers respect PHIs for insecticides more often than not. Even though both minisquash and chayote are continuously harvested vegetables—harvested every few days, which presents considerable difficulties in respecting PHIs—export farmers generally adjust their spraying times to extend the PHI by spraying immediately after harvest. In contrast to these forms of caution, farmers in both sectors commonly violate fungicide PHIs, the only aspect of their pesticide use vis-à-vis residues that corresponds with the typical farmer in developing countries.

In addition to showing how regulatory risk shapes export farmers' behavior, these data raise a number of questions about the gap between regulation and production practices. Why does pesticide use differ substantially from EPA tolerances even after exporters and the state police pesticides on export crops? Why are farmers in both sectors more cautious about insecticides than fungicides? If farmers commonly use pesticides on export crops that do not have EPA tolerances on those crops, why are pesticide residue violations not more frequent? And, how does caution with residues translate into caution by using protective gear? These questions necessitate an understanding of local context in relation to regulation.

The data point to a previously unrecognized form of caution that export farmers exercise: insecticide choice. For farmers in the study site, the most important response to regulatory risk is using pyrethroids and avoiding organophosphates, especially near and in harvest, because pyrethroids generally break down so rapidly. This is one reason why Rachel Carson (1994) suggested pyrethroids as an alternative to the more persistent organochlorines. Ironically, exporters' recommendations to use pyrethroids leads to pesticide use that is less compliant with EPA tolerances. Pyrethroids are used to avoid residue rejections, which paradoxically is *not* the same as complying with EPA tolerances. The EPA has not registered most of the commonly used pyrethroids for the crops on which export farmers use them. Only permethrin, which is one of the six pyrethroids that form the core of export farmers' insecticide regimens, has an EPA tolerance for squash and chayote (Environmental Protection Agency 2004). This highlights a broader problem for export farmers that chayote exporters brought up: squash, and especially chayote, is a minor crop in the United States, so there is little incentive for agrochemical companies to obtain EPA tolerances for them (Bischoff 1993; Boh 2003).

Embracing pyrethroids is based on locally developed knowledge of their residuality. These pesticides have relatively low PHIs on their labels, have been used for more than a decade by export farmers during harvest time, and have never caused residue rejections of produce exported from the area. Exporters interpret this to mean that the pyrethroids effectively leave no residues, and therefore they are part of the zero-residue mandate of the minivegetable exporters. The FDA regularly tests for them, and they have not caused violations in Costa Rican produce, so this interpretation is likely correct. Residue-degradation experiments show that breakdown rates are very fast (up to 50 percent per day) for cypermethrin, deltamethrin, and permethrin (Ripley et al. 2001). Thus, using pyrethroids is a useful adaptation to regulatory risk even though it does not comply with the letter of U.S. law. The question of why export farmers are far less cautious about fungicide residues remains.

A local misunderstanding of pesticide toxicity information presented on pesticide labels modifies farmers' response to regulatory risk. Many farmers interpret a pesticide's label toxicity symbol and color band (table 5.2) to indicate the level of hazard to the pesticide handler *and* the consumer (and, further, to the environment). In other words, farmers view pesticides' hazardousness along a single axis (very harmful, less harmful, etc.). Based on the assumption that U.S. residue laws would restrict the most harmful types of residues, many export farmers interpret a pesticide's color band as symbolizing its propensity to cause residue violations on produce exported to the United States. For example, a very knowledgeable export farmer who has been working to convert to organic agriculture explained the meaning of the color bands: "They come with a yellow band, red band, green band. Green is that I can spray it today and harvest tomorrow. A yellow band means that if it is sprayed today, you have to give it a space of two or three days. And a red band, it is known that if it is sprayed today, you have to wait fifteen, twenty-two days after for harvesting" (Farmer survey, September 2003).[6]

Table 5.2 Costa Rican pesticide toxicity categories, packaging symbols, and color bands

World Heath Organization acute toxicity category		Symbol on package and foldout label		Color band on package		Oral LD$_{50}$ (mg/kg)		Dermal LD$_{50}$ (mg/kg)	
		symbol	wording[a]	color	wording[a]	solid	liquid	solid	liquid
Ia	Extremely Dangerous	☣X	VERY TOXIC	red	EXTREMELY DANGEROUS	≤ 5	≤ 20	≤ 10	≤ 40
Ib	Highly Dangerous		TOXIC		HIGHLY DANGEROUS	5–50	20–200	10–100	40–400
II	Moderately Dangerous	X	HARMFUL	yellow	MODERATELY DANGEROUS	50–500	200–2,000	100–1,000	400–4,000
III	Slightly Dangerous	—	CAUTION	blue	SLIGHTLY DANGEROUS	500–2,000	2,000–3,000	> 1,000	> 4,000
IV	—	—	PRECAUTION	green	—	> 2,000	> 3,000		

[a] Original wording in Spanish, as are all elements of all labels I encountered.
Sources: Costa Rican pesticide labels collected by author and Picado Rojas and Ramírez Matamoros (1998, 372).

The problem with this interpretation is that acute toxicity, as represented visually by the color band, is a different chemical property than the amount of residue that a pesticide will leave, which is not represented graphically on the package. Some very acutely toxic pesticides such as methomyl have short PHIs because they have short half-lives, while other less acutely toxic pesticides, especially some organochlorines, persist as residues for decades and centuries. One famous example is DDT. DDT is not very acutely toxic, meaning that one can ingest a considerable amount and not experience poisoning, but it is extraordinarily persistent (leaving high amounts of residues on food) and has long-term health effects related to endocrine disruption and cancer.

This interpretation partly explains the lack of respecting fungicide PHIs, since about 80 percent of fungicides sold in the study site have a green band, while insecticides mostly have red and yellow bands. Some farmers (mis)interpret this green band to mean that fungicides do not present residue problems, and other farmers (incorrectly) consider them to be biopesticides and not harmful. While farmers' interpretation that color bands represent a pesticide's propensity to leave harmful residues may seem unreasonable, it is very logical that farmers interpret the color and wording of a pesticide's color band to mean the danger to both themselves *and* the consumer. Farmers do not receive information on the finer points of agrochemical characteristics, and the design of the Costa Rican color band symbols (see table 5.2) contributes to the problem. While red, yellow, and blue color bands all include the word "DANGEROUS," green bands do not. From this labeling scheme, it is logical (but unfortunately incorrect) to conclude that pesticides with green bands are not dangerous, and indeed, many farmers hold this view.

This misunderstanding of fungicides as not being dangerous is reinforced by the specific history of residue violations for export crops in the study site and the subsequent enforcement focus of exporters and MAG. Locally, the insecticide methamidophos is the most notorious pesticide for causing residue violations, and it has a red band. Exporters proscribe it and other red-banded highly toxic pesticides such as carbofuran and phorate as well as some yellow-banded organophosphates such as dimethoate and the organochlorine endosulfan. In contrast, green-banded pesticides have never caused residue rejections of export produce from the study site, making it easy, although incorrect, to associate a green band with low regulatory risk. For example, a nonacutely toxic fungicide, chlorothalonil, regularly causes violations of Guatemalan peas imported into the United States (Galt 2010).

This color-band interpretation and the histories of rejections due to illegal residues explain why there is a strong contrast between export farmers' caution with insecticides and caution with fungicides, even though insect and disease pest pressure are both very high in the area. Another question I posed above remains: if export farmers commonly use pesticides that do not have EPA tolerances, why are FDA rejections due to illegal pesticide residues not more frequent? This stems in large part from lax FDA testing. Further analysis that I have conducted reveals that FDA residue tests would not detect many commonly used pesticides, especially fungicides. Specifically, there is a disconnect between most of the 59 pesticides used on exported minisquash and their regulation: "Of the 25 pesticides with a squash tolerance, three would have been detected [by FDA testing], and five are exempt (for a total of 32%). Of the 22 pesticides registered in the US but without a squash tolerance, eight would have been detected (36.4%). Of the 12 pesticides not registered in the US, one would have been detected (8.3%). The ideal regulatory situation—in which a pesticide is registered, has a tolerance, and is tested for by [the] FDA or has an exemption—occurs with eight of 59 (13.6%) pesticides used." The pattern is similar for chayote's 47 pesticides: "Of the 25 pesticides with a chayote tolerance, five would have been detected, and four are exempt (for a total of 32%). Of the 17 pesticides registered in the United States but without a chayote tolerance, eight would have been detected (47.1%). Of the five pesticides without US registrations, one would have been detected (20%). The ideal regulatory situation as defined above occurs with nine of 47 (19.1%) pesticides used" (Galt 2009, 470).

In both cases, insecticides are much more likely to be detected than fungicides, since the FDA rarely employs the screens for detecting most fungicides (Galt 2011). Thus, negative regulatory feedback through residue violations has not occurred for fungicides *because the FDA is not looking for them* and therefore cannot detect them, even if their use and persistence on produce would violate U.S. tolerances.

The last disconnect in the nexus between knowledge and practice that I want to explore is farmers' use of protective gear. Farmers in the area know a great deal about the dangers of pesticides—poisoning (including

death), cancer, birth defects, brain damage, etc.—but there is no correlation between the number of pesticide dangers they know of and the
amount of protective gear worn.[7] In fact, the two are completely unre-
lated, which is a very common finding within the literature. Furthermore,
protective gear use by export farmers, who as a group are particularly
concerned about residues, is not significantly different from national
market farmers. Thus, concern about and considerable caution with pesticide *residues* does not have spillover effects in terms of farmers protecting their own bodies from exposure (Galt 2013). This suggests important
limits to the effects of regulation from afar in the study site in that it is
concerned only about processes that determine product quality, not the
overall well-being of the people and environments producing it.

Beyond the Binary of Pesticide Use/Misuse

Export farmers in the study site show considerable caution around
residues as a result of regulations in the global North being mediated
through the commodity chain. Bringing together a focus on contract
farming and the literature on pesticide use in developing countries shows
that the narrative of farmers' widespread lack of caution in their pesticide use does not hold up in the context of coordinated supply chains in
which exporters exert control over their contract farmers. Yet it is clear
that the exporters' control over farmers is far from perfect, as exporters'
own understandings of regulations are incomplete, and farmers have considerable leeway in their actual pesticide-use decisions because pesticide
policing is always partial. There are also substantial limitations to how
far regulation from afar can go to improve the quality of life and environment of the region.

Saying that farmers in developing countries misuse, abuse, or cau-
tiously use pesticides is inherently political because it can play into the
agendas of various groups debating the appropriateness of pesticide-
dependent agriculture. The agrochemical industry position, operating by
what Murray (1994) calls the safe-use paradigm, argues that providing
proper training and information to farmers can solve pesticide problems
in developing countries. Critics of the pesticide industry respond from a
different framework, which I call here the inherent-problems paradigm,

emphasizing that social, economic, climatic, and other conditions in the South make pesticide use according to the label impossible in developing countries (García 1999; Murray 1994; Wright 1986, 1990).

I suspect that there is a fear among critical researchers that describing any aspect of farmers' pesticide use as cautious plays into the safe-use paradigm. While I agree with critical perspectives that the safe-use paradigm is grossly inadequate and must be countered (Galt 2013), I also believe that we can tell accurate stories in which pesticide use by farmers in developing countries can be quite rational without supporting the narratives and goals of agrochemical promoters. Finding and explaining farmers' caution with pesticides is not to downplay pesticides' many negative effects or to paint a rosy picture about chemically dependent agriculture but is instead to acknowledge that farmers can effectively respond to regulation to tackle some pesticide problems. The story also highlights the flip side: state intervention, including regulation, is often the best way to address many pesticide problems, and state regulations work, even if imperfectly.

Disaggregating caution also allows for dismantling an unfortunate binary in pesticide research that is often mapped onto the problematic First World–Third World (North-South) divide. In the parlance of research focused on developing countries, farmers in developing countries typically "abuse" and "misuse" pesticides (e.g., Andreatta 1997; Grossman 1992; Williamson 2003), implying that farmers in the industrialized world merely "use" them. The implied distinction hides the fact that pesticide "misuse" and serious hazards exist in industrialized countries as well (Pulido and Pena 1998) and that even "proper" pesticide use, applied in accordance with the label, negatively impacts health and the environment (Nash 2006). The binary is also problematic since it prevents certain questions from being asked in certain places, thereby precluding conclusions that do not fit the assumptions of the binary and preventing the development of new theory. McCarthy (2005, 956), in discussing First World political ecology, notes "how much we can learn when we are prepared to recognize and research relations typically assumed not to exist in industrialized countries." The converse is true of Third World political ecology. In the case of pesticides, researchers focusing solely on misuse will inevitably overlook aspects of pesticide-use rationalization in developing countries since this sort of rationalization through regulation is a logical impossibility within the dominant mode of thinking and thus cannot be considered, let alone theorized.

Just as the dominant narrative of misuse hides the possibility of farmers' caution with pesticides, the almost exclusive focus of critical researchers on pesticide use in agroexport production leaves another blind spot: pesticide use in national market produce. Chapter 6 illuminates this blind spot, revealing reasons for considerable concern.

6 "It Just Goes to Kill Ticos"

On a rare cloud-free morning in 2003 I visited the farm of Manuel, who produces chayote for export and other vegetables sold on the national market. He showed me that he washes his chayote in clean wash bins dedicated solely to the task, as required by the exporter that contracts with him. As we walked across his farm and through fields of green beans and trellises of chayote, we came upon workers washing green beans for the national market, distinguishable by their larger size from the *vainica fina* (fine green beans) grown for export. The workers were washing the green beans in a fifty-gallon drum known as an *estañón,* which almost all farmers in the area use for mixing pesticides (figure 6.1). This was occurring right within the pesticide mixing area, next to pesticide bottles and other *estañones.* When I expressed shock, he said adamantly that he would never do that with his export chayote. Besides, he assured me, the *estañón* had been rinsed well since being used for pesticides.

His reason for not washing his national market green beans in his export chayote wash bin was that he sprayed the green beans with pesticides not permitted on export chayote. There is a risk that these residues would contaminate the export crop through the wash bin. The solution is using a pesticide mixing tank, rather than the export wash bins, to wash produce for his fellow Ticos (a nickname Costa Ricans give themselves because of their propensity to use a diminutive adjective form ending in "tico," such as *chiquitico* rather than *chiquito*). This double standard in the treatment of export and national market produce illustrates this chapter's focus: pesticide use on national market vegetables as it relates to residues, especially in comparison with exported vegetables.

A great deal of research focuses on pesticide residues and pesticide use as it relates to residues in developing countries. The studies follow three general contours: (1) residue testing by analytical chemists and food

Figure 6.1 Green beans for national market being washed in a pesticide mixing tank. (Source: author)

toxicologists that generally shows high levels of pesticides on national market produce (Carazo et al. 1984; Centre for Science and Environment 2006; Chang, Chen, and Fang 2005; Dogheim et al. 1990; Ip 1990); (2) assessments of the practices of pesticide use, as discussed in chapter 5, emphasizing farmers' lack of caution for both their own safety and that of consumers of their products; and (3) critical social science—generally using political ecology or political economy of agriculture approaches—that emphasizes a neocolonial environmental injustice by relating pesticide use and problems to the broader political economy and agroexports (Weir and Schapiro 1981).

The first two types of studies identified above—residue studies and those emphasizing lack of caution—generally conclude, with little supporting evidence, that pesticide problems stem from ignorance, so education or training of farmers is the best approach to correct the problem (e.g., Rodríguez Solano 1994). In contrast, the studies in the last category almost always conclude that integration of agricultural production into export markets increases pesticide use, thereby subjecting local populations and environments to high levels of pesticide contamination (Conroy, Murray, and Rosset 1996; Mo 2001; Murray 1991, 1994; Murray and Hoppin 1992; Stewart 1996; Stonich 1993; Thrupp 1991a, 1991b; Thrupp, Bergeron, and Waters 1995). This work informed by political economy remains silent on national market production, making for a substantial gap in applying a political-ecological approach to domestic produce in developing countries.

Filling this gap means confronting unanswered questions. Two are addressed here. First, why do farmers use pesticides in a manner that causes high levels of residues on produce for developing country markets? Specifically, can this be attributed to lack of education, as is often done, or more to political economic factors related to production and/or the ecological underpinnings of agriculture? Second, what are the effects of uneven pesticide residue regulations on farmers' pesticide use? In other words, how do food systems in an unequal world, expressed through different markets often with different types and strengths of pesticide regulation, affect farmers' pesticide use and the exposure of different populations fed by different market segments? These questions are important to most everyone who eats food in developing countries—the majority of the world's population.

A Regulatory Double Standard and Its Effects

When viewed globally, there is an uneven spatiality of pesticide residue regulation by national markets: the industrialized world/global North has relatively strong regulations over pesticide residues in food, while the developing world/global South lacks many regulations. I construe the term "regulation" broadly to include the creation and enforcement of rules governing behaviors, processes, and products. Chapters 4 and 5 revealed that U.S. pesticide regulation and the private standards and regulations of other industrialized countries create regulatory risk to which exporters and export farmers respond by trying to avoid illegal pesticide residues, although within the context of limited resources and imperfect understandings. In contrast, this regulatory risk is largely absent in developing countries, since enforcement of pesticide residue standards, if they exist, is minimal for produce destined for the national market (Bojacá et al. 2013; Dinham 2003; Ecobichon 2001; El Sebae 1993).

Authors in industrialized nations, especially scholars using a political economy perspective, seldom recognize or acknowledge the double standard of stronger export regulations than for the national market. Arbona (1998) provides one of the few exceptions with her work on pesticide use in Almolonga, an area in Guatemala analogous to Northern Cartago that produces potato, carrot, onion, cabbage, beet, celery, lettuce, and radish for national and Central American markets. She notes that in Almolonga, unlike in the export pea–growing areas, the idea of reducing or eliminating pesticide use through alternative agricultural practices "has been difficult to introduce. One of the main objectives of these initiatives elsewhere has been reduction of pesticide residues in food, initiatives sparked by the repeated failures and losses of Guatemalan vegetable exports interdicted on the international market. Historically, Almolongueño farmers, whose principle markets—national, Mexican, and Central American—have less stringent import regulations, have not had to introduce changes in their intensive use of pesticides" (61–62).

Similarly, while noting improvements in pesticide use and control in Senegal's agroexport sector, Williamson (2003, 13) mentions that "Senegalese public servants have pointed out the double standards existing between food safety requirements on produce exported to Europe and national capacity and monitoring of residues in local vegetables and food grains." Another example is that pesticide residue monitoring in Colombia focuses "on export products while excluding most of the food commodities" (Bojacá et al. 2013, 400). Kopper (2002, 1) observes that in Costa Rica, "since 1997, under pressure from food safety agencies in the United States of America, Costa Rica's main export market, the government and the country's large-scale export-oriented industries have been undergoing major efforts to ensure compliance with new, stricter standards on fresh produce. While these efforts have been crowned with success, much remains to be done to ensure a safer food supply to the local market." This is the regulatory double standard in developing countries: stronger regulation in industrialized nations' markets leads to pesticide-use rationalization in export production, while weaker domestic/regional regulation does not work to rationalize pesticide use in domestic market produce.

The regulatory double standard arises from similar produce requirements in capitalist markets in both North and South but much stronger regulatory power in the North. In other words, the double movement to create state power over food safety has been much stronger in the industrialized world. Although those writing in the global North from a political economy perspective rarely acknowledge it, pesticide use is generally very high on national market vegetables in developing countries. Research on these markets has concluded that produce buyers' cosmetic requirements are very important (Dinham 2003; Grossman 1998; Jungbluth 1997; Medina 1987), and the high value of horticultural crops makes it likely that they will be produced with high levels of agrochemical inputs (Fernandez-Cornejo, Jans, and Smith 1998; Galt 2008b; Gockowski and Ndoumbe 2004). Grossman (1998, 197) notes that "Vincentian farmers producing for sale in local markets are finding their customers increasingly selective concerning the quality of produce that they are willing to purchase. . . . Consequently, Vincentians strive to produce blemish-free crops for the local market by using more pesticides." Jungbluth (1997) in Thailand, Dinham (2003, 577) in Vietnam, Medina (1987, 151) in the Philippines, and Rodríguez Navarro (1983, 28) in Costa Rica all make similar observations about high pesticide use in vegetables to meet high cosmetic requirements for national markets (but see Amekawa 2013 for a contrasting situation).

Thus, like farmers in the North, many farmers in the global South must meet strict expectations for blemish-free produce. How these cosmetic standards and the expectations of perfection developed in different places is an interesting question. Freidberg (2009, 129, 144) suggests that widely circulated portraiture of fruit influenced standards of fruit beauty in the 1800s in Europe and the United States and that advertising in the 1900s meant that growers, especially in California where most shipments to the rest of the country were originating, had to strive to make fruit as pretty as the ads. A U.S. Department of Agriculture publication noted that "the appearance of an apple is everything and taste nothing. . . . [The grower] must be prepared to raise pretty red apples stuffed with cotton if his customers want them" (Holmes 1905, cited in Freidberg 2009, 144).

She notes that "eventually, grading, branding, and advertising produce became not just a California specialty but a global norm" (Freidberg 2009, 147). Costa Rican national market farmers noted this, commonly repeating the refrain that what sells produce is *que se ve,* that it looks (good).

The problem is that these high cosmetic standards are paired with considerably weaker regulatory oversight than in Northern markets, and we have seen how tempting it is for farmers to use highly effective and residual pesticides even with regulatory oversight. Consequently, national market vegetables in developing countries have serious pesticide residue problems. Residue tests done on a limited basis in developing countries reveal that pesticide residues on fresh produce are widespread and often exceed residue standards (table 6.1). These tests show that most fresh produce has detectable residues, and these very often exceed limits either of the country, of Food and Agriculture Organization standards, or of industrialized countries (various reference points are used). For example, a report from India shows that 51 percent of all food commodities had pesticide residues and that the maximum residue level (MRL) was exceeded in 20 percent, whereas these figures worldwide are 20 percent for residues and 2 percent in excess of the MRL (Agnihorti 1999, cited in Gupta 2004). Another Indian study showed that "all food commodities contain pesticide residues, and the average intake through food was estimated at 223 µg per person per day, compared with 3.8 µg in the USA" (Dinham 2003, 578). Table 6.1 does not show the level by which standards are exceeded, but it is often quite large. For example, a study of pesticide residues in Quito, Ecuador, showed that the average level of methamidophos in tomatoes was more than 15 parts per million (ppm) (Hidalgo 1980, cited in Probst et al. 1999, 54), much higher than the 1 ppm in tomatoes permitted by the U.S. Environmental Protection Agency (EPA) (Environmental Protection Agency 2004).

While there is a prominent academic discourse in the North that consumers should not worry about pesticide residues in produce (Ames, Profet, and Gold 1990; Lomborg 2001), there is also a counterdiscourse arguing that these residues over a lifetime can pose real risks (Baker et al. 2002; Benbrook 2002) or at least should be avoided out of precaution (Pollan 2006, 179). Absent from both of these discourses is available

Table 6.1 Pesticide residues on fresh produce in developing countries, various years

Product	Samples	Percentage with residues detected	Percentage with illegal residues	Place	Source
Vegetables	1000	—	35 % (OC)[a]	Hyderabad market, India	1
Leafy vegetables	300	—	100 % (BHC)[b]	Mysore, India	1
Horticultural crops	—	—	40%	Kenya	1
Vegetables	—	—	>50%	Sri Lanka	2
Vegetables	—	—	>50%	Togo	2
Vegetables	—	37% (OC)[a]	—	Thailand	3
Cowpea	—	—	10%	Thailand	3
Tangerine	—	—	10%	Thailand	3
Kale	—	—	20%	Thailand	3
Chile peppers	—	—	87.5% (malathion)[c]	Egypt	4
Cereals, pulses, milk, eggs, meat and vegetables	—	53%	34%	Egypt	4
Produce	—	—	2–5% (OC&OP)[d]	farmers' markets, Brazil	4
Produce	—	—	41%	supermarkets, São Paulo, Brazil	4
Produce (11 types)	—	—	80%	supermarkets, São Paulo, Brazil	4
Lettuce	70	—	100%	Nicaragua	5
Cabbage	140	—	71%	Nicaragua	5
Chilitoma	85	—	65%	Nicaragua	5
Watermelon	117	—	41%	Nicaragua	5
Tomato	200	—	27%	Nicaragua	5
Potato	25	—	8%	Nicaragua	5
Eggplant	6	100%	0%	Delhi, India	6

continued

Table 6.1 Continued

Product	Samples	Percentage with residues detected	Percentage with illegal residues	Place	Source
Cabbage	7	100%	0%	Delhi, India	6
Tomato	7	100%	71%	Delhi, India	6
Cauliflower	7	100%	43%	Delhi, India	6
Chile peppers	5	100%	20%	Delhi, India	6
Okra	7	100%	29%	Delhi, India	6
Mustard	5	100%	40%	Delhi, India	6

[a] OC = organochlorines, including DDT, endrin, aldrin, etc.
[b] BHC is an organochlorine insecticide, also known as lindane.
[c] Malathion is an organophosphate insecticide.
[d] OP = organophosphates.
Sources: (1) Bull (1982, 57–58); (2) Schwab (1995, 42); (3) Palakool (1995) cited in Jungbluth (1997); (4) Dinham (1993, 57, 90, 169); (5) Anonymous (1998, 30); (6) Mukherjee (2003, 270).

evidence that high residue levels can and sometimes do cause consumer poisonings. The organophosphates and carbamates are of particular concern. In some cases food-borne pesticide residues kill people, although these cases generally result from inadvertent contamination and not from pesticide applications in agricultural production.[1] More common are hospitalizations due to food poisoning from pesticide residues,[2] although there are few registrations of these types of poisonings, and they are rarely discovered, let alone reported in the literature (Wu et al. 2001). Reported symptoms of pesticide poisoning from residues on food are nausea, vomiting, chills, sweating, dizziness, abdominal pain, diarrhea, and headache, which—while symptomatic of pesticide poisoning in general—are easily mistaken by consumers and doctors alike for symptoms of other types of food poisoning, especially from bacterial agents (Green et al. 1987). Only with "careful clinical and epidemiological investigation" can the origins of food poisoning outbreaks due to pesticides be identified (Wu et al. 2001, 333–34).

The recent documented cases in the medical literature come from Malaysia, Taiwan, and Hong Kong. Methamidophos—the very acutely toxic, systemic organophosphate insecticide that featured prominently in chapters 4 and 5—was the culprit in all cases. In September 1991, eleven people in Malaysia were hospitalized due to food poisoning from methamidophos residues on the leafy green vegetable *sayur manis,* which had methamidophos residue levels five hundred times above the permitted level (Dinham 1993, 179). In investigating four pesticide residue poisoning cases in Taiwan in 2001, researchers found that leftover vegetables contained extremely high levels of methamidophos: 225 ppm in sweet potato, 110 ppm in the leafy green *Gynura bicolor,* and 26.3 ppm in red cabbage (Wu et al. 2001, 335). For a reference point, the U.S. maximum residue level (or tolerance) for methamidophos ranges from 0.1 to 1 ppm (Environmental Protection Agency 2004, 404). Taiwan banned the use of methamidophos on most vegetables in 1994, but Wu et al. (2001, 336) report that "because it is cheap and highly potent, the illegal use of methamidophos is still common among farmers."

Poisonings from methamidophos residues on produce were once common in Hong Kong, where physicians began watching for it specifically. In 1988, Hong Kong experienced 312 documented outbreaks of consumer poisonings from methamidophos residues (figure 6.2). Hong Kong was able to decrease the frequency of these types of poisonings—down to 18 outbreaks in 1995—by banning methamidophos in 1988 and by spot-testing imported produce, most of which is Chinese, for residues before it entered the country (Chan 2001). While the Hong Kong case suggests useful corrective measures, it remains more common that poisonings from high levels of pesticide residues on food are not diagnosed and recorded as such but instead pass as bacterial food poisoning.

Other cases of acute illness linked to pesticide residues exist but are not in the medical literature. Dinham (1993, 34) reports that in Egypt, cases of illness such as diarrhea have been attributed to pesticide-contaminated fruit and vegetables for decades. This accompanied the rise of vegetable cultivation in plastic houses and tunnels, which results in "heavy residues from frequent applications of pesticides, without observing the safe interval between the final spray and harvesting" (56). The culprit insecticides in Egypt are mostly organophosphates and carbamates: dimethoate, methamidophos, methomyl, and monocrotophos. Sudan also has had residues cause pesticide poisonings. In the 1990–1991 cotton

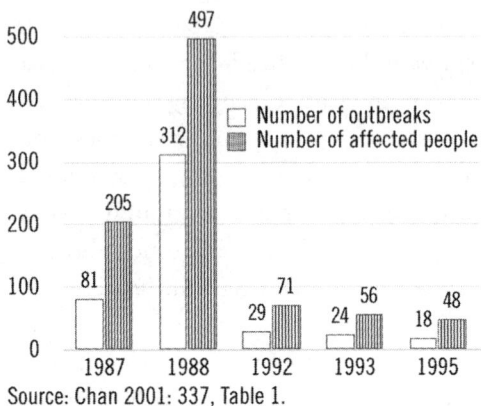

Source: Chan 2001: 337, Table 1.

Figure 6.2   Vegetable-borne methamidophos poisonings in Hong Kong.

season, forty-five pesticide poisoning cases were admitted to hospitals in the cotton-growing provinces of Gezira and Blue Nile. Surprisingly, twenty-five cases were caused by eating pesticide-contaminated vegetables or wheat (46–47). In Paraguay, there is concern about the increasing numbers of pesticide poisonings from consumers eating aboveground produce such as cabbage, lettuce, and tomatoes (128).

These incidents and other limited evidence suggest that the burden of residues in developing countries is considerably higher than in industrialized countries and has observable consequences. Higher levels of intake lead to higher body burdens. For example, one study estimated that the average daily intake of HCH and DDT in India was 115 and 48 μg per person, respectively, which is much greater than intake in industrialized nations (Kannan et al. 1992). As one would expect, very high levels of pesticides are found in samples of human milk, fat, and blood in India (Bhatnagar 2001, cited in Gupta 2004). Links between pesticide residues and other negative health effects of pesticides—immune system suppression, nervous system damage, cancer, and birth defects—have not been explored and are very difficult to show in noncontrolled studies.

Shifting our focus to Costa Rica, pesticide residue studies show that the rate at which Costa Rican national market produce violates residue standards is quite high, from 9 percent to 30 percent (García 1997; Rodríguez Solano 1994; Valverde G., Carazo Rojas, and Araya Rojas 2001). The most recent study of vegetables from Northern Cartago had a violation rate of 17 percent, although a substantial portion of pesticides used in the area could not be detected with the analytical chemistry methods used (Fournier L. et al. 2010). These rates are substantially higher than the violation rate of Costa Rican produce imported into the United States—4.4 percent for all Costa Rican fresh vegetables imported from 1996 to 2006—and also of U.S. domestic produce consumed in the United States, which was 1.6 percent from 1996 to 2006 (Galt 2010). While these rates cannot be directly compared due to difference in sampling laws, residue analyses conducted, and violation rates due to illegal or excessive residues are far greater in Costa Rican produce for the national market than for exported produce and for internally consumed produce in the North.

The next section examines the specific market segments that exist in Costa Rica and the different strengths of pesticide residue regulation within them. Before analyzing them, I want to note that the double standard is likely being complicated by the rapid expansion of supermarket chains in developing countries (Balsevich et al. 2003; Berdegué et al. 2005; Dugger 2004; Reardon and Barrett 2000), which might be creating an increase in contract farming relationships for coordinated supply chains for the national market. Local supermarkets as market outlets now "equal or exceed the importance of non-traditional exports in the Central American [fresh fruit and vegetable] sector," although the effects of this transformation are much less studied (Berdegué et al. 2005, 256). In Costa Rica, about 50 percent of food retail sales went through 227 supermarkets in 2002 (Berdegué et al. 2005, 256–57). As in industrialized countries, consumers in developing countries are increasingly concerned about pesticide residues, and some want assurance that the produce they eat will not poison them (Díaz-Knauf et al. 1993; Dinham 2003; Reardon et al. 2001). An important question for Central America is how supermarkets' expansion "is affecting, if at all, the quality and safety standards of the [fresh fruits and vegetables] sold and consumed in the region" (Berdegué et al. 2005, 256).

National Markets and Pesticide Residue Policing

Three market channels are relevant to pesticide residues in Costa Rica: the open domestic market, the controlled domestic market, and the export market. The open national market consists of farmers selling their produce to intermediaries or directly to consumers at *ferias* (farmers' markets). The intermediaries sell to retailers such as supermarkets and produce stands and to various businesses that supply the Costa Rican food industry. A recent study found that these food industry businesses are mostly small, with 93 percent having fewer than one hundred employees, and that 59 percent of Costa Rican food industries had no raw material specifications (Kopper 2002, 2–3). The vast majority of these companies have not adopted food safety standards because of their cost and because most consumers "are not aware of their right to safe foods and are mainly interested in purchasing items at the lowest possible price" (2). For national market intermediaries and farmers supplying them, there is little pressure to comply with pesticide residue standards.[3] Valverde G., Carazo Rojas, and Araya Rojas (2001, 82) note that "the intermediaries that provide the produce markets and the roving truck farmer are not held accountable directly if contamination is detected in their products." Similarly, ferias place no demands on farmers in terms of monitoring pesticide residues. Like in most developing countries, the majority of produce in the Costa Rican national market faces minimal enforced demands to comply with Costa Rican pesticide residue standards.

The controlled national market refers to the Blue Seal (Sello Azul) program, a Costa Rican agrifood standard. The Blue Seal program is run by the Phytosanitary Department (known simply as Sanidad Vegetal) within the Ministry of Agriculture and Livestock (MAG) and by supermarkets with the goal of certifying produce that complies with pesticide tolerances (Kopper 2002). The program is voluntary and is designed mostly for supermarkets or large intermediaries that supply supermarkets, reflecting a broader trend toward private standard regulation in agrifood systems (Fuchs, Kalfagianni, and Arentsen 2009; Le Heron and Roche 1999; Marsden, Flynn, and Harrison 1999; Murdoch, Marsden, and Banks 2000). To receive the seal, the company must comply with the standards for six months and regularly supply a determined number of samples for laboratory analysis. These samples should include the names and addresses of farmers so that MAG can advise them on proper pesticide use (Castillo Nieto 1999).[4] Businesses certified under the Blue Seal program can label their produce with a sticker saying "Business with control of pesticide residues" (Castillo Nieto 1999, 17).

The first company to receive the Blue Seal was Hortifruti División Vegetales, owned by Corporación de Compañías Agroindustriales de la Corporación de Supermercados Unidos (Grupo Más por Menos). As part of this program, Hortifruti "invests resources in random tests with the goal of detecting agricultural contamination among [the produce of] its providers" (Valverde G., Carazo Rojas, and Araya Rojas 2001, 82). Hortifruti is by far the largest fresh fruit and vegetable supplier in Costa Rica. The company has informal contracts with farmers and acts similarly to exporters, relying on "careful selection of growers and then the maintenance of a stable relationship" and providing farmers with "stable access to an attractive and growing market, at prices that are close to but usually a bit above the wholesale market, plus technical assistance, and for the small farmers, input credit" (Berdegué et al. 2005, 265). Hortifruti also conducts pesticide residue tests through MAG's Phytosanitary Department lab, which cost $200 for the farmer (266). If violations are found in "either the pesticide or E. coli fronts, this is used to orient the technical assistance and training activities of their field staff rather than to signal delisting of the supplier" (266).

While the Blue Seal signals greater control over pesticide residues in vegetables, its current implementation appears to allow for the skirting of monitoring because companies can choose the produce to submit and also because the Blue Seal program does not contact every farmer involved. Viewed skeptically, the Blue Seal program is a way of differentiating produce sold in supermarkets from that sold in farmers' markets, implying that it has fewer or lower pesticide residues and is of higher quality (Berdegué et al. 2005). Valverde G., Carazo Rojas, and Araya Rojas's (2001) residue study points to the problems with this view, since pesticide residues examined were present with equal frequency in produce sold at farmers' markets and supermarkets.[5] However, a more recent study of pesticide use in the Pacayas-Plantón area (see figure 0.1) shows that in contrast to open-market farmers, "farmers that sell to supermarket chains use very little pesticide and use more biopesticides and biofertilizers, without affecting their productivity or profitability" (Fournier L. et al. 2010, 25). Thus, it is still unclear precisely how the controlled national market influences pesticide use and residues.

As shown in chapter 5, the export market places demands on farmers to comply with pesticide residue standards. The U.S., Canadian, and European regulatory agencies spot-test produce for pesticide residues. While less than 1 percent of produce is tested, the main exporters in Northern Cartago and the Ujarrás Valley all take pesticide residue violations seriously, since the cost of violations is the loss of a shipment worth $10,000 or more. Some have experienced violations in the past, and now all use policing mechanisms to exert control over pesticide use. Exporters generally proscribe organochlorine and organophosphate pesticides and suggest the use of pyrethroids and biopesticides in their stead.

While lax compared to its export markets, Costa Rica's regulatory framework for pesticides has improved recently, but monitoring and enforcement in the open national market are still minimal. As of 1990, Costa Rica had established pesticide tolerances for only twenty-five pesticides on strawberries and seven on dried cacao (García 1990). In 1997, a large number of tolerances were established in the official government register, La Gaceta (1997). Costa Rican law also dictates that the foods in the standard market basket be tested for pesticide residues. MAG's Phytosanitary Department is responsible for conducting residues tests on national market produce with its laboratory, which began monitoring in 1991 (Castillo Nieto 1999, 17). This was spurred in part by investigative reporting in the country that highlighted extremely lax enforcement, widespread pesticide misuse, and high levels of residues, especially in vegetables (Barquero S. 1990; Barquero S. and Navarro 1990a, 1990b; Navarro and Barquero 1990).

In 1999, MAG's Phytosanitary Department was conducting fifty tests per week for the national market, imported goods, and exports. Each residue test cost MAG's Phytosanitary Department 45,000 colones, equivalent to $158 at the time. Some exporters "turn to this service to present proof of health [innocuousness] to those countries where they send their products" (Barquero S. 1999, 25A). Although this testing exists, these residue tests for national market produce have little impact on the open national market channels. The vast majority of farmers and produce buyers participating in my study were unaware of sanctions against farmers who sold produce with high levels of residues on the Costa Rican market. Farmers almost universally agreed that they can use pesticides with impunity when selling on the open national market. Thus, even though there is a regulatory framework on which to base monitoring and enforcement, these are relatively lax, likely because of two decades of structural adjustment and continued high foreign debt levels that have sapped budgets for government programs generally.

The Logic of Using Residual and Toxic Insecticides

In terms of residues on food, an important choice that vegetable farmers make is choosing between the less residual insecticides (such as pyrethroids and newer biopesticides) and the generally more persistent and acutely toxic organophosphates and carbamates, as noted in chapter 5. Without regulation, farmers tend to choose the organophosphates because they are more effective per monetary unit spent. Table 6.2 shows the average agrochemical cost (not including labor and other costs) of an insecticide application in the study site by chemical class and active ingredient. While it is not immediately apparent that organophosphates are the most effective insecticides per monetary unit, one must consider how persistent the insecticide is on and in the plant. High preharvest intervals (PHIs) generally indicate pesticides that break down slowly, while low PHIs show rapid breakdown.

The problem with the pyrethroids from a production standpoint is that most degrade to below effective levels in a few days, leaving the plants unprotected from insects. This is their benefit to the consumer and the farmer seeking to avoid residues: they leave very low levels of residues after only a few days, commonly below the detection limit of commonly used residue analysis equipment.

In contrast, many organophosphates, especially the systemic ones such as methamidophos, have much longer half-lives and persist at high enough levels to kill insects often for weeks after application. Some export farmers explained the trade-off of having to spray the synthetic pyrethroids more frequently than the organophosphates to achieve the same level of control. For example, one farmer when asked if export requirements for low residues affected his production practices responded that "The only

Table 6.2 Cost of insecticides per hectare, Northern Cartago, July 2003

Chemical class / active ingredient	Toxicity class	Oral LD$_{50}$ (mg/kg)	Average cost/ha (US$)[a]	PHI
Organophosphates (OP)				
chlorpyrifos	II	135–163	$7.47	20
diazinon	II	300–400	$13.08	10
dimethoate	II	290–325	$9.44	14–21
methamidophos	Ib	20	$8.53	10–21
	Average	465.3	$9.63	15.75
Organochlorines (OC)				
endosulfan	II	70	$10.93	2
	Average	70	$10.93	2
Pyrethroids (P)				
cyfluthrin	II	869–1,271	$13.33	14
cypermethrin	II	247	$5.19	0–15
deltamethrin	IV	128–5,000	$12.85	8
lambda-cyhalothrin	II	56–79	$6.98	1
permethrin	II, IV	4,000	$10.41	0
	Average	1,343.1	$9.75	5.08
Synthetic insecticides with new modes of action				
imidacloprid	III	450	$57.94	21
teflubenzuron	IV	> 5,000	$13.84	15
thiamethoxam	IV	1563	$22.47	3
	Average	2,337.7	$31.41	13
Botanical, microbial, & organic insecticides				
avermectin	II	10	$98.91	0
Bacillus thuringiensis	IV	> 5,000	$23.18	0
potassium salt, oleic acid	IV	> 5,000	$23.55	0
spinosad	IV	3,738	$37.27	3
	Average	3,437	$45.73	0.75

[a] Averages were calculated by determining the cost per hectare of each formulation and package size of the pesticide available at the store, for both the minimum and maximum dose recommended on the label. A 2003 exchange rate of CR¢395/US$1 was used.
Sources: Instituto Regional de Estudios en Sustancias Tóxicas (1999); pesticide labels; and analysis of General Inventory Report from Insumos Agropecuarios ADICO, Cipreses, Cartago Province, Costa Rica, July 28, 2003.

thing it does is you have to apply a little more, shorten the cycle of applications. Let's say, if I spray with residual pesticides [organophosphates] every eight days, with the nonresiduals [pyrethroids] I have to spray every five days" (Farmer survey, September 2003). In a conversation with Rodrigo, a farmer who used to produce entirely for the national market and had switched to the export market, I was trying to understand what many farmers and exporters communicated as the continuing temptation to use methamidophos as opposed to the less residual and newer pyrethroids.

RG: Is it because methamidophos is cheaper?

RODRIGO: It is not cheap, but rather the effect on the plant lasts longer than the products I use, the pyrethroids. The pyrethroids, they have a shorter duration on the plant. Meanwhile, methamidophos, if you apply methamidophos, you know the plant is protected for fifteen, twenty-two days.

RG: And still there are people who use methamidophos in chayote?

RODRIGO: Yes. It is prohibited. It is prohibited, but people spray it on. People spray it because in chayote it is the product that one [traditionally] uses. For example, cypermethrin [a synthetic pyrethroid], in chayote, cypermethrin doesn't do a thing for you.

RG: So what do you use in chayote? Do you have to use methamidophos still?

RODRIGO: No, because I use a higher dose of cypermethrin.

RG: Does it work like that?

RODRIGO: It works like that. But you have to spray it more frequently.

RG: So it costs a little more.

RODRIGO: It costs a little more.

RG: And if you didn't have a fixed price [as a result of the policy of two minivegetable exporters in the area], you would always be trying to have the lowest costs.

RODRIGO: That's right. (Farmer interview, December 2003)

There are a few other references to this connection in the literature. Opondo (2000, 38), writing on French beans produced in Kenya for export to Europe, notes that "the high cost of environmentally sound agrochemicals[6] causes a lot of farmers to avoid using some of those recommended, and to continue to use proscribed chemicals. But this can only be in the short term because testing for the maximum residue regulation has already begun in Europe, and produce not found adhering to this regulation will be rejected." Arbona (1998) also makes this connection although at the national level. As the export farmer above describes, it is not just price but price relative to how well the agrochemicals work and how long they remain effective in the field.

This logic helps explain why many studies find that national market farmers tend to use more residual pesticides, especially organophosphates and methamidophos, that can leave high levels of residues that cause consumer poisonings (e.g., Abeysekera 1988; Yen, Bekele, and Kalloo 1999). Alone, however, this does not adequately explain high residues on national market produce, since farmers could apply these pesticides and wait until they dissipate before harvest. High levels result from not waiting long enough, that is, not respecting the PHI.

PHI is often not respected for two main reasons: the frequency of harvest of many crops and the variability of prices that farmers receive. Considering the first reason, many vegetable crops by their biological nature are continuously harvested every few days or every week for weeks or months on end. These include botanical fruits used as vegetables, such as chayote, cucumber, eggplant, green beans, peas, peppers, squash, and tomato. Once these are in harvest, the maximum time a farmer can wait between a spray and a harvest is the period between harvests, which ranges from a couple of days to a week depending on the crop and variety. Thus, when highly residual organophosphates are used during this continuous harvest period, high levels of residues will result, especially during the first few harvests after the spray.

Residue decline studies show what happens to methamidophos residue levels depending on the number of applications and the PHI. For a residue decline study conducted on cucumbers grown in greenhouses in Spain, after only one application of methamidophos it takes almost three weeks for residues to drop below 0.1 ppm, the level that is detected by most residue tests. In contrast, the regular use of methamidophos causes residues to accumulate in the plant, since it is a systemic insecticide and takes several weeks to break down (figure 6.3). The frequency of application in figure 6.3 is once every two weeks. By the fifth application, considerable residues have built up in the cucumbers over the course of the production cycle. Tests on pepper and tomato display the same pattern

(Aguilar-del Real, Valverde-García, and Camacho-Ferre 1999). Concerning the final levels of methamidophos, the authors of the study note that "these data seem to indicate the existence of a certain long-term accumulative effect for this pesticide. These results are in agreement with the high residue levels of methamidophos sometimes found by regulatory agencies in residue monitoring" (3357–58). A Costa Rican study of methamidophos residues in lettuce and tomato also found very high residues. Lettuce sprayed with six applications of methamidophos every two to three weeks and harvested with a one-day PHI and a three-day PHI had residues of 12.79 ppm and 9.74 ppm, respectively (Carazo et al. 1984, 943), while tomatoes harvested eight days after the last application had residues of 4.92 ppm (942).

Turning to the second reason that PHI is not respected, price variability can also lead to residue problems in vegetables that are harvested once or twice at the end of the production cycle, such as broccoli, cabbage, celery, lettuce, and sweet corn. National market vegetable prices fluctuate strongly at annual, weekly, and even daily intervals. Edelman (1999, 96)

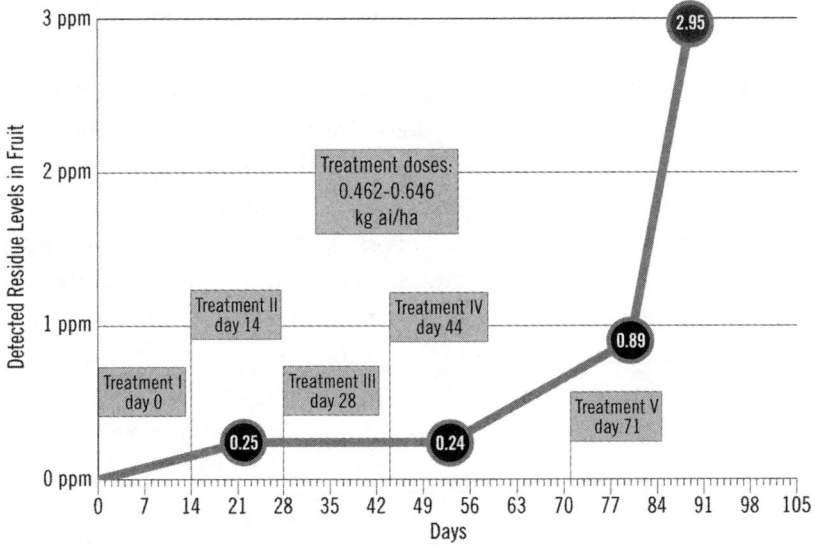

Source: Aguilar-del Real et al. 1999: 3356 & 3358, Tables 1 & 6.

Figure 6.3 Levels of methamidophos residue in cucumber during and following five consecutive applications.

describes the Costa Rican national market prices for vegetables as "chaotic"; onions and potato prices, "the Cartago agriculturalists' principle crops, unlike those for coffee and basic grains, were [and continue to be] highly unstable." In the interview with Rodrigo, he explained the effects of this on production strategy: with national market vegetables you have to always keep input costs as low as possible, since "you're gambling with the price on the market" (Farmer interview, December 2003). This often means preference for organophosphates to keep total agrochemical costs lower. The connection to residues, farmers explained to me, is that near harvest time a rapid increase in price can mean that less conscientious farmers, or those in a cost/price squeeze who have recently sprayed, will harvest to take advantage of the price. The recently sprayed produce, likely with high levels of residues, will be consumed in the next day or two. Washing does not help with methamidophos residues, as it is a systemic insecticide that is translocated throughout the plant, including all edible portions. This situation explains why some farmers in the area refuse to eat cabbage, since they say that the pesticide use and harvesting practices of their neighbors result in very high levels of residues.

We can therefore identify a triad of causes behind high organophosphate and carbamate residues on vegetables: (1) higher efficacy per unit price (with significant pest pressure assumed), together with (2) the biological characteristic of continuously harvested vegetables, and (3) rapid price fluctuations in vegetable markets. Together these causes mean that in markets with little regulation, farmers will generally prefer the organophosphates, and harvests will often occur without enough time passing between spraying and harvesting. This explains the high levels of residues often found in developing countries. It has very little to do with farmer ignorance and much more to do with the intersection of farm households' survival in capitalist markets and the ecological aspects of vegetable production, including crop biology and susceptibility to pests.

Pesticide Use by Market Segment

What are the effects of different levels of pesticide residue regulation on farmers' pesticide use as it relates to residues? During the open-ended section of the farmer survey, I asked farmers about the requirements of the national market and the export market. The pattern of observations were remarkably consistent. A common understanding is noted by Carlos, a smallholder who sells his produce in farmers' markets: "If you sell to the national market, in reality no one says to you, 'Look, I'm going to do a residue test.' Here the agricultural market is open; no one is put to work to see if residues are high, or anything else, no" (Farmer survey, April 2003). Other farmers noted that laws existed but were not strongly enforced. Juan, who produces vegetables for export and had previously been oriented only to the national market, noted about the national market, "Well, there are requirements, but they are not put into practice. The national market, MAG has a person working in the Phytosanitary Department. They do tests, but they are not very effective, because they do them very sporadically" (Farmer survey, May 2003).

This contrasts with export farmers' views of export market requirements. Juan noted the requirements concerning high cosmetic standards and residues: "good formation of the product so that they are sent perfect, almost. . . . And all free of residues, free of residues. This is one of the most restrictive limitations they have. . . . This clearly has a large cost for us and [is done] as carefully as possible. I always want to learn much more to use products that are less toxic every day, and, if it is possible, nothing" (Farmer survey, May 2003). While not all export farmers share his thoughts on stopping synthetic pesticide use, his views of residue requirements in the export market are typical, as discussed in chapter 5.

In contrast to rather shared understandings of open national and export markets, farmers' views of pesticide residue policing within the controlled national market, as represented by HortiFruti, are divided. Some farmers said that HortiFruti randomly tests for residues and pays for the tests, while others were adamant that farmers had to bear the costs of residue testing and that it was only done if the farmer wanted it done. Others reported that HortiFruti exercised no control over their pesticide use, only caring about the amount and appearance. The manager of one vegetable exporting business, which has participated in the Blue Seal program, is skeptical of its efficacy and implementation so far: "Basically, there's no follow through on it. For example, Hortifruti has tons and tons of different producers and they have the Blue Seal, but nobody does the *registro*. So it doesn't mean anything" (Manager interview, April 2003).

Thus, it is difficult to make sense of exactly what kind of effect the Blue Seal program has on farmers' production practices since there is little shared understanding of the program.

With these different levels of residue enforcement between different markets, especially the differences in the open national market and the export market, we would expect differences in farmers' use of the more residual and highly toxic organophosphate and carbamate insecticides and perhaps in other pesticide classes as well. Table 6.3 compares groups of pesticides used on three crops—chayote, green bean, and squash—produced for different markets. The statistical tests compare only pesticide use on crops for the export market and the open domestic market, since the sample sizes for the controlled national market are too small for statistical inference.

Pesticide use for the different markets shares a number of features. First, pyrethroids are the insecticide group with the highest levels of use across all crops and markets, followed generally by the organophosphates and carbamates. Second, the dithiocarbamates (e.g., mancozeb) are an important fungicide group on all crops. Third, the substituted benzenes, especially chlorothalonil, and inorganic fungicides such as sulfur and copper are the other important fungicide groups.

Table 6.3 also reveals significant differences in insecticide use between the open national market and the export market. First, production for the open national market relies on higher use of organophosphate and carbamate insecticides. This difference is significant across all three crops. Second, for green bean and squash, open national market farmers rely significantly less on purchased botanical, microbial, and organic insecticides—including avermectin, *Bacillus thuringiensis,* and spinosad—than export-oriented farmers. Third, squash and green bean farmers who sell to the open national market rely significantly more on the older and more persistent organochlorine insecticides than do export farmers of the same crop. Fourth, squash and chayote farmers selling to the open national market rely significantly less (at the 10 percent level) on newer insecticides with new modes of action (Nauen and Bretschneider 2002) than do export squash and chayote farmers.

There are also some differences in fungicide use between the markets. First, green bean export farmers rely more on botanical fungicides than do national market green bean farmers. Second, there are many differences in fungicide use between squash production for different markets. National market squash farmers rely significantly more on dithiocarbamates and inorganics, while export squash farmers use more antibiotics, botanicals, and other types of synthetic fungicides.

The sample sizes shown in table 6.3 of farmers producing for the controlled domestic market are too small to draw strong conclusions, but the table suggests that their insecticide use is closer to that of export farmers than open national market farmers. As with export crops, pyrethroids are emphasized over the use of organophosphates. It is interesting to note that some of these marketing arrangements were made with farmers already producing for export.

Another way to draw pesticide comparisons is using their negative health consequences. The Pesticide Action Network (PAN), an activist and research organization, created the PAN Bad Actor classification to differentiate between pesticides known to cause problems and those that are likely to be less harmful (Kegley, Orme, and Neumeister 2000). A pesticide qualifies as a bad actor if it is any of the following: a highly acute toxin according to the World Health Organization, the EPA, or the U.S. National Toxicology Program; a known or probable carcinogen according to the EPA; a reproductive or developmental toxin listed in California's Proposition 65 list; a cholinesterase inhibitor according to the Material Safety Data Sheet, the California Department of Pesticide Regulation, or the PAN staff's evaluation of chemical structure; or a known groundwater contaminant (Orme and Kegley 2004). Despite its creation by an activist organization, the PAN Bad Actor category is relatively conservative. Not included are suspected endocrine disruptors, possible carcinogens, and suspected groundwater contaminants.

Table 6.4 compares the use of bad actor pesticides by market segment for chayote, green bean, and squash. In all crops, the total number of pesticide doses per crop cycle is significantly lower for export produce. The total number of bad actor pesticides follows the same pattern: significantly more bad actor doses are used on produce going to the open national market. When one controls for the difference in total doses by looking at bad actor pesticides as a percentage of all doses, their use is lower for all export crops, but the difference is significant only for

Table 6.3 Groups of insecticides and fungicides used by market segment, Northern Cartago and the Ujarrás Valley, 2003–4

	Chayote				Green Bean				Squash			
	National		Export		National		Export		National		Export	
	Open (n=13)	Controlled (n=3)[a]	(n=20)	p[b]	Open (n=22)	Controlled (n=3)[a]	(n=11)	p[b]	Open (n=33)	Controlled (n=7)[a]	(n=28)	p[b]
Insecticide groups[c]												
Pyrethroids (P)	52%	82%	63%	0.28	56%	76%	55%	0.26	53%	50%	41%	0.11
Organophosphates & carbamates (OP/C)	40%	4%	20%	0.05	26%	12%	18%	0.03	33%	19%	17%	0.00
New modes of action (BU/C-N/Neo/Py/T)[d]	5%	0%	14%	0.10	6%	0%	5%	0.26	4%	15%	17%	0.07
Organochlorines (OC)	0%	0%	1%	0.16	7%	0%	0%	0.09	6%	0%	0%	0.03
Other synthetic insecticides (N/AI)	0%	0%	1%	0.16	5%	0%	6%	0.26	4%	0%	3%	0.30
Botanicals, microbials, & organics (Bot/Mic/So)	4%	14%	1%	0.35	0%	12%	16%	0.01	0%	15%	22%	0.00

Fungicide groups[c]

Fungicide groups[c]												
Dithiocarbamates (DC)	41%	64%	42%	0.48	50%	20%	35%	0.21	45%	42%	35%	0.02
Inorganics (In)	42%	36%	45%	0.29	9%	0%	6%	0.12	24%	6%	5%	0.01
Substituted benzenes (SB)	11%	0%	8%	0.07	16%	64%	21%	0.12	19%	27%	14%	0.29
Azoles and benzimidazoles (A/B)	5%	0%	3%	0.31	13%	8%	10%	0.23	8%	15%	15%	0.20
Antibiotics (AB)	1%	0%	2%	0.17	3%	4%	7%	0.34	0%	6%	10%	0.00
Other fungicides (An/CM/M/OtC/Th/Un/X)	0%	0%	0%	—	8%	4%	14%	0.19	5%	2%	16%	0.00
Botanicals (Bot)	0%	0%	0%	—	1%	0%	7%	0.08	0%	2%	5%	0.00

[a] Data on controlled national market production are de-emphasized with gray shading because sample sizes are small and are not included in the statistical tests.

[b] Two-tailed t-tests assuming unequal variance, comparing open national and export categories.

[c] Table A.1 shows the pesticide active ingredients included in different pesticide classes as denoted by the abbreviations in parentheses.

[d] As defined by Nauen and Bretschneider (2002): acetamiprid, chlorfenapyr, cyromazine, diflubenzuron, fipronil, imidacloprid, novaluron, teflubenzuron, and thiamethoxam.

Source: Author's farmer surveys, 2003–2004.

Table 6.4 Use of bad actor pesticides per crop cycle by market segment, Northern Cartago and the Ujarrás Valley, 2003–2004

	Chayote				Green Bean				Squash			
	Domestic		Export	p^b	Domestic		Export	p^b	Domestic		Export	p^b
	Open	Controlled[a]			Open	Controlled[a]			Open	Controlled[a]		
	(n=13)	(n=3)	(n=20)		(n=22)	(n=3)	(n=11)		(n=33)	(n=7)	(n=28)	
Total pesticide doses	104.4	126.7	72.7	0.02	23.6	15.3	18.9	0.09	46.11	25.86	30.05	0.01
Bad actor doses:												
Highly toxic	19.5	2.2	10.9	0.12	3.5	1.3	1.8	0.07	7.2	2.0	3.5	0.01
Cholinesterase inhibitor	20.8	2.2	8.4	0.03	2.1	0.7	0.4	0.02	5.3	1.4	1.8	0.01
Carcinogen	18.6	0.0	16.4	0.34	7.8	6.0	5.5	0.19	12.2	7.4	9.6	0.14
Developmental or reproductive toxin	33.1	45.2	20.2	0.04	9.9	2.7	5.2	0.02	15.8	10.7	12.6	0.21
Groundwater contaminant	0.0	0.0	0.0	—	0.0	0.0	0.1	0.33	0.0	0.3	0.0	—
Bad Actor total (any of the above criteria)	55.0	47.3	33.9	0.04	15.5	9.3	10.3	0.04	29.0	17.9	17.8	0.01

Bad Actor doses as percentage of all doses:

Highly toxic	18%	1%	14%	0.21	15%	9%	9%	0.08	18%	13%	11%	0.03
Cholinesterase inhibitor	19%	1%	11%	0.06	10%	4%	6%	0.09	14%	10%	5%	0.01
Carcinogen	17%	0%	21%	0.24	28%	39%	35%	0.24	26%	25%	29%	0.22
Developmental or reproductive toxin	30%	36%	27%	0.29	37%	19%	25%	0.03	33%	35%	39%	0.27
Groundwater contaminant	0%	0%	0%	—	0%	0%	0%	0.23	0%	1%	0%	—
Bad Actor total (any of the above criteria)	51%	38%	44%	0.16	61%	63%	58%	0.34	64%	63%	55%	0.06

[a] Data on controlled national market production are de-emphasized with gray shading because sample sizes are small and are not included in the statistical tests.

[b] Two-tailed t-tests assuming unequal variance, comparing open domestic and export categories.

Source: Author's farmer surveys, 2003–2004.

squash. More specific differences are revealed in the "highly toxic" and "cholinesterase inhibitor" categories. Organophosphates and carbamates are cholinesterase inhibitors, and most are highly toxic. Thus, the differences in those categories largely reflect the differences between insecticide classes used, as shown in table 6.3. The significantly lower number of doses of reproductive and/or developmental toxins in chayote and green bean for export largely reflects lower overall fungicide use, since many common fungicides in the area—e.g., mancozeb, maneb, propineb—are reproductive and/or developmental toxins. The lack of difference in this category for squash produced for the different markets in part reflects export squash farmers' adoption of newer fungicides such as thiophanate-methyl that appear to have the same carcinogenic and reproductive/developmental toxin properties as the older fungicides they replace. It is important to note that no information on carcinogenesis, developmental or reproductive toxicity, or groundwater contamination potential is available to farmers in the area. Pesticide labels only communicate acute toxicity via the color band and provide PHI, registration information, and emergency procedures in small print.

Looking in more detail at the case of squash grown for the different market segments (see appendix 2, figure A.2), it is obvious that farmers' respect of the recommended dose is actually quite high in both market segments. On average, only five of forty-eight (10.4 percent) of the active ingredients used for open national market squash were applied in excess of the dose on the label. Of these, four are from only one farmer. For exported squash, on average only three of fifty-nine (5.1 percent) were applied as overdoses, and all result from one farmer's use.

In contrast to general respect for dose, PHIs for most pesticides sprayed on squash are commonly violated due to the continual harvests, high pest and disease pressures, and an almost exclusive reliance on agrochemicals for pest control. Sixty-nine percent of the average minimum PHIs violate the label requirements in open national market pesticide use. In contrast, 36 percent of the average minimum PHIs for the export market are in violation of required PHIs. National market farmers are almost twice as likely to be in violation of PHI requirements.

Regulatory compliance regarding which pesticides can be used on squash is also relatively low in the open national market. While all pesticides are registered for some use in Costa Rica, of the pesticide active ingredients used by open national market squash farmers, only 29 percent are registered specifically for use on squash in Costa Rica. An additional 27 percent have tolerances set for at least one other cucurbit—chayote, cucumber, or melon—but not squash. On the other end of compliance—residue tests by regulatory agencies—there is also a strong disconnect. Residue tests performed by MAG's Phytosanitary Department (Rodríguez Solano 1994) would find only pyrethroids, organophosphates, organochlorines and only one of twenty fungicides used on national market squash. The regulatory situation in export market squash has a similar level of disconnect, as noted in chapter 5.

The general pattern from these comparisons shows less caution by national market farmers vis-à-vis residues. This lower level of caution is evident in less respect for PHIs, higher dependence upon and more frequent use of organophosphates and carbamates, and less use of botanical, microbial, and organic insecticides, which avoid residue problems but are very expensive relative to synthetic agrochemicals. In addition to considering general patterns, there are also a number of illustrative cases of open national market farmers' pesticide use.

While not representing the average condition, these cases, described below, also demonstrate the double standard of pesticide residue regulation and its manifestations in farmers' management. The first case is Raúl, who aims production at the national market with the *negro* variety of chayote and at the export market with the *quelite* variety of chayote. He has two different systems of insecticide use for the different markets. For national market chayote, he uses mostly organophosphates—chlorpyrifos and a pesticide that is a combination of dimethoate and cypermethrin—and waits for a minimum of two days between spraying and harvesting. The labels specify twenty days for chlorpyrifos and seven to fifteen days for the combination pesticide, so he is constantly violating these PHIs. For his export chayote, he uses only pyrethroids—cypermethrin, deltamethrin, lambda-cyhalothrin, and permethrin—and also waits two days between spraying and harvesting. Most of these pyrethroids have low PHIs, so these practices would translate into higher pesticide residues on the national market produce.

The residue double standard is also evident in the practices of some farmers who generally produce only for export. Vidal, who produces chayote for export, said that he sometimes applies residual, proscribed pesticides such as methamidophos or carbofuran for very intense pest outbreaks in export chayote. I asked what he did with the harvest, since chayote is generally harvested every few days for a period of nine months or more:

VIDAL: Well, when it is in full harvest, it is always in harvest, always in harvest. It doesn't stop. For this reason one is very careful, because of the danger you run. If it comes out contaminated, they close the export market to you. They won't receive chayote for a period of time, until . . . methamidophos or the poison has already passed its cycle.

RG: In harvest, it doesn't go for export?

RUBÉN: [interrupting the conversation] It goes to kill Ticos.

VIDAL: [nodding his head in agreement] It just goes to kill Ticos. (Farmer interview, November 2003)

In the above conversation, Rubén, an export squash farmer, had introduced me to his neighbor Vidal, a chayote farmer, and had then gone to work on his own farm. Rubén returned when he was done with the task and sat down beside Vidal and me under the tin roof shed in order to get out of the rain that had started to fall. Rubén had heard only a minute or so of our conversation but understood where it was going. Without Rubén's return, it might have been hard for Vidal to admit that immediately after the application of very residual and toxic organophosphate or carbamate insecticides, he sells his chayote to the national market instead of the export market. One other chayote export farmer reported using the same practice during the survey, and another chayote farmer, Manuel, whose story started this chapter, pointed out that he uses methamidophos but only on his green beans for national market.

The two export chayote farmers who admitted to selling highly contaminated produce to the national market account for 10 percent of the export chayote farmers in the survey. It is important to note that they volunteered this information to me and that I did not specifically probe other farmers to try to get them to admit this practice. I suspect that more export chayote farmers might follow a similar strategy, although few are willing to talk about it due to the delicate nature of the subject and the unethical nature of the action. Thus, for what is likely an important minority of export chayote farmers, the national market serves as an outlet for chayote that export farmers and exporters consider to be too contaminated by pesticide residues to be safe to export. This is similar to produce of second-tier cosmetic quality being routed to the national market, as is commonly done in export sectors, but likely has much greater implications for the health of consumers.

Unlike some export chayote farmers, all export squash farmers aim their squash production for the export market. Because the export and national markets for their produce are tightly controlled by the two exporters, export squash farmers do not have the myriad national market outlets that chayote farmers can use, so they sell it all to either exporter. These firms grade the produce, with the lower grades going to various channels in the controlled national market and sometimes to ferias. Managers of the exporters insisted that this is determined entirely by cosmetic and size requirements and that they do not consider or test for residues in making the decision. Thus, the squash sold in the controlled national market by the two exporters are produced in the same manner as those squash that are exported (see also Breslin 1996).

The difference in potential residues on squash for the different markets comes from some national market farmers who grow squash varieties different from those exported and engage in pesticide use that is completely proscribed by exporters, especially the use of methamidophos during harvest. As noted above, consecutive use of organophosphates, especially methamidophos during the harvest phase, can result in high concentration of the pesticides in the plant and its fruit. For example, Miguel, a farmer who grows squash for national market, uses methamidophos once every two weeks on his squash during harvest time. With his twice-weekly harvests, there is a maximum of four days between application and harvest, although the label requires at least twenty-one days. Similarly, Cristóbal sprays methamidophos once a week on his national market squash for the entire cycle, for a total of eighteen applications. He reported waiting three to four days between the spray and the harvest and also reported that the individual squashes require about a month to develop fully. This means that each squash as it develops receives four doses of methamidophos at the high rate he uses, 1.143 kilograms (kg) of active ingredient (ai) per hectare (ha); the recommended dose on the label is 0.3 to 0.9 kg ai/ha. With the practices of both of these farmers, residue levels likely build up to levels higher than those shown in figure 6.3, since applications are more frequent. Another national market farmer uses en-dosulfan, an organochlorine insecticide, on his squash every week. As with frequent methamidophos applications, these frequent applications would likely lead to very high residue levels in his squash. These three farmers, whose squash is likely to be highly contaminated with pesticide residues that pose a consumer poisoning risk, represent 11.5 percent of the national market squash farmers in the survey. Most other national market squash farmers use highly residual pesticides in harvest but not as frequently.

With these difference in farmers' pesticide-use practices between markets, exported squash and chayote likely have lower insecticide residue levels on average than squash and chayote produced exclusively for the open national market. From a consumer poisoning standpoint, there are likely extremely high levels of residues in about 10 percent of national market cucurbits if my surveys are representative.

Explaining Pesticide Residue Problems

Remedies for high levels of pesticide residues on produce for national markets in developing countries typically focus on farmer education. Gupta (2004, 89) notes that "there is every reason to properly educate farmers for judicious use of pesticides." Another study discusses legislation and accompanying regulation as a short-term solution to pesticide residue problems but concludes that "the long-term solution to pesticide problems is education" (Ecobichon 2001, 32). The only available report from Costa Rica's national pesticide residue laboratory repeats the common assumption of those who conduct residue analyses by suggesting visits to farms where violative residues have been found, "with the goal of establishing the reason for excessive pesticide residues and collaborating with the producer, checking application equipment and the dose of pesticides used, evaluating pest problems, and training them in good use of pesticides" (Rodríguez Solano 1994, 50). The assumption of these corrective actions is that violative residues are a technical problem that can be fixed at the individual level by training the farmer and reviewing her or his application equipment.

A technical explanation is inadequate because it does not deal with the "simple reproduction squeeze" (Watts 1983a) of the small vegetable farmer in Costa Rica facing declining or fluctuating prices. As simple commodity producers with almost no subsistence activities, Costa Rican vegetable farmers depend on the prices they receive for their produce for basic household reproduction. This situation creates a temptation to use the more effective and residual organophosphates and carbamates to keep production costs low while still protecting the crop. Pesticides that leave lower levels of residues are generally more expensive or less effective over time, and the unpredictable nature of prices in the national market means that farmers must keep costs low in case harvests occur during times of low farm-gate prices, which frequently happens. High pest pressure, the biological nature of the continuously harvested vegetables, and the high effectiveness of the more residual pesticides intersect with the simple re-production squeeze and a lack of regulatory oversight to result in high levels of residues of organophosphates and carbamates, a sign that they offer the most cost-effective, short-term "solution" to the pest problem that farmers face. Thus, a fine-grained analysis of pesticide use as done here—including farmers' decisions about pesticide type, dose, and PHI in the context of effectiveness per monetary unit spent, pest pressures and biological characteristics of crops, and forms of control in various markets—reveals a different and better explanation for residue problems.

A fine-grained analysis also offers important nuances and qualifications to the view of widespread lack of caution. Most farmers, even in the open national market, take into account the residue question, so it is not just regulatory risk that makes farmers cautious about pesticide use. Even open national market farmers who face little to no effective regulatory pressure to rationalize pesticide use still exercise some cautions with residues. This is due to a basic concern for the well-being of others, since most farmers in the area believe that pesticide residues can harm the health of those who consume their produce. However, a handful of national market farmers admitted to not caring about residues or to being economically driven enough to spray and harvest soon afterward without regard for the consumer. One said that he intentionally sprays the day before harvest so that the produce buyer will not have any problems with the produce rotting. This kind of admission was not common, and many farmers who professed their own caution around residues noted that other farmers who spray immediately before harvest would not admit to doing so.

Most farmers have a knowledge of residues and profess some obligation toward the consumer. Grossman's (1992, 1998) work similarly points to St. Vincentian farmers' concern about exposing their families and national market consumers to toxic pesticide residues. Concern about consumer exposure to residues is the first layer of the causes that create caution over residues and pertains to all market segments in Costa Rican vegetable production. Export farmers have the same concern, but layered on top of it are exporters' enforcement mechanisms (see chapter 4), making it so that export farmers exercise more caution around residues than national market farmers.

The implications of overturning the farmer ignorance thesis with the political ecological understanding from this chapter are profound. A guaranteed or less variable farm-gate price for produce is extremely important—it allows farmers to plan for and afford the higher input costs of the least residual pesticides because they do not have to constantly be aiming for the lowest costs while still protecting their crop. Thus, the relatively high levels of export squash and export green bean farmers' use of organic, microbial, and botanical insecticides can be understood in the context of their specific market conditions: (1) their farm-gate price is fixed and has moved incrementally upward over the years, and (2) their exporters are adamant about avoiding residues on exported produce. As Rodrigo, an export squash farmer, pointed out, "You have a guaranteed price; you know that it is a fixed price. So, you can take care not to spray residual products" (Farmer interview, December 2003). Export chayote farmers do not receive a fixed price, and it has even declined over time in the last few years, but they are still pressured by exporters to avoid residues. As a consequence of the declining price, they do not use the least residual organic, microbial, and botanical insecticides, as other export farmers do, but because of some control by exporters they instead rely more on the pyrethroids and avoid organophosphates.

These examples show how promoting farmers' economic well-being is a crucial underpinning in the transition to less damaging forms of pest control and that it must be paired with larger forces pushing in, or promoting, that direction. I explore this and other implications next in the conclusion.

Conclusion   A Green Agriculture for the

Green Republic?

I met Jorge Martínez in his potato field near his retail plant nursery in Pacayas, where he grows starts of broccoli, cabbage, cauliflower, and other crops for farmers in the area. He showed me around his farm, noting the many organic production methods: California earthworms create worm compost and a foliar spray of worm compost tea, compost from plant material and manure builds soil organic matter, yellow sticky traps control leaf miners (*Liriomyza* spp.) and whitefly (*Bemisia tabaci*). He hardly sprays insecticides, and the potato variety he is breeding for resistance to late blight allows him to drastically cut his fungicide use. After the survey I calculated the pesticide intensity for his potato production, which sits in the middle of the cloud belt: 12 kilograms (kg) of active ingredient (ai) per hectare (ha) per cycle. This is many times lower than the average in the cloud belt—63.8 kg ai/ha/cycle—and is even below the average of farmers who grow in the highest elevations of Volcán Irazú and Volcán Turrialba, where growing conditions are most favorable (see chapter 2).

Jorge is deeply committed to the principles and practices of organic production, and his nursery business provides a stable income that acts as a solid economic foundation upon which he can experiment with his potato production. But of more than seventy potato farmers I spoke with, he is the only one who is this dedicated to organic agriculture. Most farmers whose family incomes depend entirely on their sales on the vegetable market do not have much economic stability. Quite logically, they are anxious about the loss of their crop, since it is a considerable investment of capital and labor and is usually the basis for their entire income for that growing cycle. This short-term economic dependence on high-investment crops with the possibility of a high return and high loss in addition to the very high levels of pest and disease pressure in the area creates the extremely high levels of pesticide use in the region. Most farmers will need to be guaranteed a stable income to move toward less pesticide- and fertilizer-intensive practices or must have very strong intrinsic motivation to try to change, since it is economically very risky and goes against the cultural norms of spraying in the area (Bell 2004; Galt 2013).[1]

Jorge's foundation of economic stability from his nursery business is akin to the set price of the minivegetables and the ways in which these set prices support farmers in adopting alternative practices. The relatively strong regulation in the U.S. and European markets and this stable price act together to shape and support minivegetable export farmers' pest control practices, which as we have seen are less pesticide intensive than those of national market farmers and rely to a considerable extent on biological pesticides and alternative agricultural methods to reduce reliance on agrochemicals. These findings show that high pesticide use is not inherent in the area, although there are strong social structures that buttress pesticide dependence in powerful ways. This chapter uses an understanding of these structures—detailed immediately below—to offer possible solutions that could be implemented by the Costa Rican state.

Political Ecological Insights

Political ecology has insisted for a quarter century that broader political economic processes strongly influence land users' decisions and their environmental outcomes. Instead of acting as a predeterminate lens, political ecology as an approach prioritizes certain sets of relationships and typically uses inductive logic (abstracting from empirics to theory) to understand them rather than saying precisely what these relationships will be like in one context or across contexts. My analytical approach has been to use abstraction (see Lewontin and Levins 2007, 149) to tease out the political economic and ecological processes and relationships operating at various scales while maintaining that they are part of a complex whole.

This approach, combined with detailed and rigorous fieldwork, creates considerable analytical purchase on complex phenomena such as pesticide use and agrifood system governance. That many of my findings go against common critical understandings of pesticides and views of localized food systems suggests the importance of fieldwork with attention to geographical and historical context. The findings also show that when the approach is broad enough to capture the myriad relationships that structure most socioecological systems, and empirically grounded and critical of its own knowledge claims, a political-ecological approach can find unexpected relationships and create new understandings. My study was multimethod for data collection and analysis because the research objects of the research are so diverse (see appendix 1). Thus, I have combined historical, structural, qualitative, quantitative, comparative, and spatial analysis of primary and secondary data.

Many of this book's key insights were possible only through simultaneous attention to political economic relationships, including the real-world and multiscalar workings of agrarian capitalism and its markets shaped by extraeconomic regulation, and to the ecological conditions that shape and are shaped by agricultural production. Consider the main points of the previous chapters concerning the socioecological dynamics of agrarian capitalism in the study site:

—Pesticides allow for the spatial extension of the production of crops into regions not well suited for them, thereby creating (1) expanded conditions for capital accumulation (especially for off-farm capital), (2) regional crop lock-in whereby farmers generally must grow high-value crops to make payments to land that increases in value, and (3) ecological contradictions as pests overcome the temporary control afforded by pesticides.

—Organisms, ecological processes, and their spatial and scalar configurations intersect with the circuits of off-farm and on-farm capital to play a large role in socioeconomic differentiation in agriculture, as resource-rich farmers access better growing environments to reduce pesticide use and increase profit rates while resource-poor farmers face a faster and more powerful pesticide treadmill.

—Vegetable production for the national market in Costa Rica (and likely other areas in developing countries) has long been heavily dependent upon synthetic pesticides since their introductions in the 1940s and 1950s, often to a surprising extent and sometimes decades before the introduction of agroexport relations for fresh produce.

—Exporters take export market regulations seriously when noncompliance affects them economically, and they therefore mediate export market requirements—albeit imperfectly—by directly shaping the production process through various mechanisms of control in their contract relationships with farmers.

—Export farmers weigh regulatory risk and loss of market access from noncompliance against short-term economic gains from using pesticides prohibited by exporters and generally err on the side of caution, especially when they are supported economically with a fixed price for their produce.

—A regulatory double standard means that farmers often use pesticides prohibited for export crops on their national market crops because they face severe cost/price squeezes in the national market and because the state lacks strong regulatory oversight.

These findings should shift stories of pesticide use onto a more complex terrain, one that recognizes it as fundamentally shaped by agrarian capitalism—organized locally, nationally, and transnationally—and upon which a regulatory double movement acts in geographically uneven ways. Placing agrarian capitalism more centrally in discussions of the pesticide problem means recognizing that it is not the scale of markets but instead the development of capitalist markets—both for farm-produced commodities and for farm inputs as commodities—that precipitate widespread and intense pesticide use, high-value regional crop lock-in (wherein producers need to produce high-value crops to afford land prices), and the treadmill of production of which the accelerating pesticide treadmill is an important part. Export markets set these dynamics in motion in many cases, but so does national market expansion. It is the core dynamics of agrarian capitalism, not export markets per se, that create many of the pest and pesticide problems experienced in capital-intensive production systems. We can no longer conceive of national and local market production as environmentally benign, as this is a serious oversight stemming in large part from the scalar-trap conceptualization (Born and Purcell 2006; Brown and Purcell 2005).

We have also seen that states—acting nationally and transnationally—attempt to shape agricultural production through regulatory interventions, and these efforts are more and less successful. A large remaining question is whether pesticide regulations acting from afar in other contexts have similar effects in other agroexport sectors. Since the topics of research in this book were mostly abandoned by agrifood researchers, only a few studies answer the question. One has shown that European Union standards pressure pesticide rationalization among export vegetable producers in Kenya (Okello and Okello 2010), while another study of Thai pummelos shows that produce exported to eastern Asia is more heavily sprayed than national market pummelos, in part because of different quality demands in which national market consumers prefer blemished fruit (Amekawa 2013). Further detailed analyses are needed to examine how these processes are unfolding in other commodity chains and how they are impacting producing regions.

Another source of information for examining the applicability of the results of my research is secondary data on pesticide residue violations, which show that many agroexport commodity chains have not experienced pesticide residue violations recently. That some transnational vegetable commodity chains rarely violate U.S. regulations suggests that actors in these vegetable commodity chains have adapted to extraeconomic demands of the U.S. market or, alternatively, that oversight by the U.S. Food and Drug Administration of these commodity chains is particularly lax. For example, violation rates from illegal residues on vegetables from Italy, the Netherlands, Thailand, Belgium, and Taiwan are lower than the average violation rates of vegetables produced within the United States for domestic consumption (Galt 2010). In contrast, noncompliance persists in some places. Methamidophos—the pesticide described by farmers and exporters in Costa Rica as one that will greatly increase production yet will cause violations—is by far the most commonly found illegal pesticide residue on produce imported into the United States (Galt 2010). This suggests that the intersections of local ecologies and broader political economy identified above pertain to vegetable production systems around the world. The contradictions structuring methamidophos use are not just a local phenomenon. These intersections of agrarian capitalism, pests, and transformed local ecologies span thousands of commodity chains and hundreds of countries, yet there is little research that looks at these issues systematically.

Although synthetic pesticides are sprayed over vast swaths of Earth's agricultural land, political ecologists and allied researchers have largely abandoned research on the role of pesticides in conventional agriculture and conventional commodity chains. Many of the researchers who were interested in agroexports in the 1980s and early 1990s shifted focus in the mid-1990s, turning to fair trade and organic commodity chains, certification generally, and, more recently, local agrifood systems and alternative food networks. These changes in agrifood systems are certainly important and need further research, but this shift by researchers has been disproportional relative to the size of these alternative systems. This leaves a considerable gap in the understanding of conventional agriculture around the world and the ways that its dynamics might best be shifted. I have sought to partially fill this gap, but my research and that of the handful of other researchers who have continued work on pesticides (Harrison 2008; Harrison 2006; Jansen 2008) and/or conventional agroexports (Amekawa 2013; Fischer and Benson 2006; Freidberg 2004; Hamilton and Fischer 2003; Okello and Okello 2010) cannot hope to understand the thousands of different conventional commodity chains around the world, let alone their economic, health, and environmental effects.

Highlighting and helping to address the continuing pesticide problems in conventional agriculture can influence national and transnational food systems, an important effort given the billions of people who are dependent upon them. There is much to be done in terms of both research and policy. Below I apply the insights from the research to propose a number of solutions to many of Costa Rica's pesticide problems. My focus is largely at the national level and in the relatively short term, since that is where Costa Rican policy can have the greatest impact. I leave out discussions of reforming or reconstituting agrarian capitalism and detailed policy recommendations on supporting the expansion of agroecologically oriented agriculture, as these are taken on elsewhere (Buttel 2006; Kremen, Iles, and Bacon 2012).

Reducing Pesticide Use in the World's Most Pesticide-Intensive Country

A sad irony of pesticide use is the unintentional displacement of disease from one organism (the crop plant) to another (humans). The net effect of pesticide use is to temporarily save crop plants from insect pests and plant diseases, but a heavy dependence on pesticides that cause cancer, birth defects, compromised nervous and immune systems, and other serious problems leading to human diseases, even if not as surely as the chemicals kill pests or prevent them from damaging plants. Pesticides can also be extremely ecologically disruptive through their toxic effects on important nonhuman organisms on the farm and in other contexts. These effects on humans and nonhumans is unintentional and hard to document yet no less real. Another sad irony, identified in chapter 1, is that pesticide use becomes a necessary practice for individual farmers' competitiveness, but this paradoxically undermines the long-term well-being of farmers, rural populations, and agroecosystems and does not guarantee their continued economic well-being.

In the study site, pesticide risks are borne in part because of export production but mostly because of national market crop production, which still dominates cultivated land. Thus, farmers, farmworkers, and other citizens of Costa Rica's "vegetable basket" pay the price—in lives, health, and environmental quality—for the vegetable production that supplies the entire Costa Rican market. Recent studies show environmental contamination from pesticides in groundwater and surface waters of the area at high enough levels to impact aquatic ecosystems and biodiversity (Fournier L. et al. 2010). Meanwhile, the Costa Rican public, when consuming vegetable crops produced nationally, facing high burdens of pesticide residues, in some cases at levels high enough to cause consumer poisonings, as has been documented elsewhere. The only actors guaranteed to win in this current configuration of the food system are the agrochemical suppliers, including transnational chemical corporations, local formulation companies, and input retailers. While these companies profit, those who pay the price are the Costa Rican rural population and environment, other citizens who eat pesticide-laden national market produce, and the health care system.

Regulation is required to promote a more just outcome in Costa Rica. The effects of regulation documented here demonstrate that understanding and ultimately shaping agrifood governance in conventional commodity chains should be an important focus for critical researchers and progressive citizen groups. My findings strongly support a major argument of economic geographers and sociologists, environmentalists, and all but the most ardent market fundamentalists: markets, especially food markets with multiple links in the commodity chain, must be regulated by the state to serve the public good. Pesticide regulation in conventional food production is an absolute necessity, since economic incentives to use pesticides at the farm level commonly conflict with consumer and environmental health. Cutting costs in food production—an imperative of capitalist competition—often harms consumers and on-farm and off-farm environments, making regulation of agrifood systems an absolutely necessary. Indeed, this book shows that in transnational commodity chains, regulation *works,* even if imperfectly, and can be improved through better understanding of its workings in real-world contexts. Similar types of regulation for the Costa Rican national market are necessary to protect Ticos.

What hope remains for the residents of Northern Cartago and the Ujarrás Valley for reducing the contamination of their living spaces, the threat of chronic and dreaded illnesses such as cancer, and harm to future generations in the form of a contaminated environment (groundwater, soils, etc.)? What about justice for the general Costa Rican public that consumes the often heavily pesticide-laden produce from the country's "vegetable basket" while less residual pesticides are used on exports? The pesticide treadmill and degradation and marginalization are not inevitable. Instead, they are the result of specific dynamics of agrarian capitalism and can be reduced or solved through attention to and intervention in the processes causing them. Although not many farmers are reducing pesticide use through adopting organic methods, the handful doing so show that farmers can greatly decrease their pesticide use and maintain yields in the area, even in the cloud belt (see also Fournier L. et al. 2010). These pioneering farmers and their efforts need economic and technical support, and it should not be just for export that these efforts are made. Their efforts can also act as demonstrations for farmers interested in reducing their pesticide use.

Costa Rica has much to be proud of, including its stable democracy, a universal health care system with health outcomes comparable to or better than industrialized nations, and its successes in environmental conservation through protected areas. Yet one of the country's largest challenges, the pesticide problem, has been inadequately addressed to date because of a lack of political will. Costa Rica's agricultural production is the most pesticide intensive in the world. Within the country, the area of

Northern Cartago and the Ujarrás Valley is likely the most pesticide intensive area—pesticides are used more heavily per hectare than even the notorious banana-growing regions—with common yearlong vegetable rotations being ten times more heavily sprayed than the national average. The regulations of export markets have helped somewhat to rationalize export farmers' pesticide use, but this has done and will do little to reduce and rationalize the pesticide use of national market farmers.

Costa Rica has the opportunity to excel in not just environmental protection through setting aside protected areas but also in environmental health for its human population and its agriculture. Moving away from synthetic pesticide dependence should be a priority for policy makers and the public. In addition to poisonings and cancer, other dangers, not documented in the study site but shown in other places, include increased risk of fetal death from environmental exposure to certain pesticides in rural areas (Bell, Hertz-Picciotto, and Beaumont 2001), impairment of cognitive and motor abilities in children living in pesticide-dependent vegetable-production areas (Guillette et al. 1998), and earlier onset of breast development in young girls (Guillette et al. 2006). These effects are likely just the tip of the iceberg.

Tackling pesticide problems for the national market will require a multifaceted approach, including

—strong regulatory oversight, with the goals of eliminating poisonings, cancer risks, and residue problems, by learning from the successes and problems of other regulatory regimes, including export market regulation;

—more stable market prices for national market produce, which will allow for farmers to better plan their input strategies and conform to enhanced national market regulation; and

—participatory farming systems research and farmer field schools to prioritize and develop agroecological solutions that can reduce the extremely high levels of pesticide use.

This approach will take considerable state effort and pressure from citizens, but it is an effort that will reap numerous benefits in terms of the well-being of farmers, farmworkers, rural residents, consumers, and environments as well as the long-term productivity of agriculture.

Regulatory oversight should focus first on the organophosphates and carbamates. In Costa Rica, these two chemical classes caused at least 34.3 percent of all poisonings in 1996 (Leveridge 1998, 43), while they made up 15.4 percent of the volume of imported pesticides (Chaverri 1999, 17). More broadly, they are responsible for the majority of pesticide poisonings, especially among farmworkers, in developing countries (Ecobichon 2001; Gupta 2004; Roberts et al. 2003). Similarly, paraquat, an herbicide, and endosulfan, an organochlorine insecticide, cause a large number of poisonings (Wesseling, Corriols, and Bravo 2005). One-third of farmers in my survey who spray pesticides have been poisoned previously by pesticides. Figures for farmworkers are likely much higher, especially since farmers are more likely to spray more heavily when it is others doing the work (Galt 2008c). Training in protective gear use is largely ineffective because it ignores considerable constraints that farmers and farmworkers face, so outright bans on the most dangerous pesticides are the most effective intervention (Galt 2013; Murray and Taylor 2000).

Carcinogenic pesticides should also receive special scrutiny. Costa Rican rural areas have a higher risk of cancer than the country's urban areas, as opposed to the typical relationship of higher cancer rates in urban areas in most countries; this is likely caused by high levels of agricultural pesticide use (Wesseling et al. 1999). Northern Cartago also has very elevated stomach cancer rates, which most farmers attribute to their heavy agrochemical dependence. Additionally, childhood cancers have been linked to parental pesticide exposure in Costa Rica (Monge et al. 2007). Despite these problems, Costa Rica has not yet ratified an important international treaty—the C139 Occupational Cancer Convention of the International Labour Organization that went into effect in 1976—that seeks to prevent and control occupational hazards of carcinogens. Ratifying C139 and following its recommendations, especially the first two of "periodic listing of carcinogenic substances that would be prohibited or made subject to authorization and control" and "replacement of carcinogenic substance," is a vital first step toward preventing more cancer cases (Monge et al. 2007, 302).

Policy prescriptions should take the form of direct controls, such as bans and other restrictions, and the use of price incentives, such as imposing ad valorem taxes on agrochemicals (Fernandez-Cornejo, Jans, and Smith 1998). Bans should target the most acutely toxic and carcinogenic pesticides and would be a significant first step. Economically, this strategy makes the most sense because it creates the greatest benefit for the least cost. In Ontario, Canada, more than 70 percent of the "total value associated with the reduction in environmental risk is due to the reduction of high-risk pesticide use" (Brethour and Weersink 2001, 226). Central American health officials have already created and agreed on a list of pesticides to ban. These are the twelve that are most responsible for mortality and morbidity across Central America: aldicarb, carbofuran, chlorpyrifos, endosulfan, ethioprophos, methyl parathion, methamidophos, methomyl, monocrotophos, paraquat, phosphine, and terbufos (Nieto Z. 2001). It would be wise and in line with its environmental reputation for Costa Rica to follow these recommendations.

Methamidophos in particular causes a large number of farmer, farmworker, and schoolchildren poisonings in addition to being the most problematic pesticide for residue-caused rejections in the export market. Other territories, notably Hong Kong, have banned the use of methamidophos and tested for it in imported produce, which led to a significant reduction in the poisoning of consumers who ate contaminated produce. Fungicides, many of which are carcinogens and teratogens (substances that cause birth defects), also need much more attention from Central American states, as noted above. Implementing the ban agreed upon by Central American health officials, including banning methamidophos, and having health agencies thoroughly review fungicides to create a list of carcinogenic pesticides to ban in a similar manner would benefit the health of children, farmers, farmworkers, and consumers and would follow Costa Rica's precedent of banning highly toxic pesticides such as aldicarb. Importantly, these bans would be done for the benefit of the national population rather than as an economic response to regulation imposed from afar.

The banning of the most problematic pesticides should be complemented with a pesticide tax. In Costa Rica, pesticides are exempt from all taxes (Agne, Waibel, and Ramírez 1999). The problem for pesticide taxes is that generally, "farmers' responsiveness to price changes for pesticides is small in the short run and small to moderate in the long run. This means that substantial taxes would be needed to achieve moderate reductions in pesticide use, particularly in the short run" (Fernandez-Cornejo, Jans, and Smith 1998, 477). Yet pesticide taxes make sense from a green economy point of view, since they will help to internalize the costs of the negative externalities created by pesticide use that are borne by society and the environment (Hawken 1993; Pimentel 2009). Not taxing pesticides that have negative externalities is an indirect subsidy of these products (Agne, Waibel, and Ramírez 1999) that contributes to unfair outcomes, as noted throughout this book. These pesticide taxes should be phased in over a number of years with a regular schedule so that farmers know they are coming and can adjust accordingly, and the taxes should be used to fund programs that help farmers adopt alternative forms of pest control.

These changes in pesticide regulations will help reduce the residue risks that Costa Rican consumers face. However, strong residue controls in the national market will need to be implemented, since from an ethical perspective the Costa Rican state should extend at least the same level of protection to its own citizens as those it has helped implement for consumers in the global North. If this is done at the level of retail outlets, these retailers will be forced to create coordinated supply chains—much like those of the export sectors—with pesticide policing mechanisms to influence farmers' pesticide-use decision making. The export sectors studied here already serve as a model of how this can be done successfully; however, the case of Guatemala shows that produce buyers' use of strong control mechanisms can backfire if adequate financial support, especially in the form of fair prices for produce, is not provided to growers (Galt 2010). Thus, fair pricing must be paired with these control efforts.

To complement increased state regulation of residues, there should be more state control over vegetable quantities and prices. This will reduce the rapid swings in produce prices that commonly entice farmers to harvest soon after spraying if prices rise rapidly. Much can be learnt from U.S. Depression-era price support systems, which have another role of supporting family farms generally (Cochrane and Runge 1992; Levins 2000) in addition to being a support that can help reduce residue problems. As export market relationships show, farmers facing a known market with set prices will be less likely to try to minimize costs by using the most residual pesticides, although this fundamentally needs to be coupled with regulatory enforcement of residue standards.

Agroecological alternatives to the riskiest synthetic pesticides need to be identified and promoted. The revenues from pesticide taxes should be funneled into research and extension efforts focused on alternative production systems as well as providing a type of insurance or other financial support for farmers who want to become organically certified. These pesticide-reduction research efforts in the form of agroecological participatory research should be focused on the most pesticide-intensive crops, especially potato, tomato, onion, carrot, cabbage, and cauliflower produced for the national market (Barnett et al. 1996). The benefit of targeting potato is that it is the most commonly planted annual crop in Northern Cartago and is also one of the most pesticide intensive. It is also important to learn from other successful pesticide-reduction efforts, most notably Indonesia's dramatic lowering of pesticide use in rice production (Thiers 1997) and Ontario's recent 50 percent pesticide-use reduction (Brethour and Weersink 2001; Surgeoner and Roberts 1993), and from farmer-to-farmer movements and related learning techniques (Bunch 1982; Holt-Giménez 2006).

Market interventions should play an important role here by supporting farmers who implement sustainable agriculture techniques. This type of financial support will be important in Northern Cartago and the Ujarrás Valley, as the risks that farmers face in their transition away from pesticides are large, the labor costs might be very high, and farm households' livelihoods and economic well-being are at stake.

State support for a large-scale conversion to more sustainable agriculture—including organic agriculture—would be an important extension and validation of Costa Rica's green reputation and would decrease the multitude of problems that heavy pesticide dependence creates. Unlike many countries in which pesticide problems are severe, Costa Rica cannot use resource poverty as an excuse to ignore the health of its agricultural production systems, ecosystems, and rural residents but should instead make the political and economic investments needed to become a sustainable agriculture success story.

Appendix 1   Study Methodology

My hope in detailing my methodology and precise methods is for others to use them and improve upon them as they seek to understand the ways in which agrarian capitalism, food system regulation, and ecological processes shape and are shaped by farmers' production practices and livelihoods. Also, by making agrochemical use more understandable and by providing one example of how to research it, I hope that more scholars will be interested in this important topic and pursue that interest. These methods, especially combining them, require some discipline and perseverance, but I believe they are worth the effort since they reveal new insights not possible through a single method alone.

Most of the primary data presented in this book were collected from a variety of methods used during ethnographic fieldwork in Costa Rica for a couple of months in 2000 and for thirteen months from 2003 to 2004. Data on pesticide use is not regularly available in most of the world, necessitating primary social science research methods to collect information on farmers' pesticide use. Much of the data originates from a face-to-face farmer survey that I conducted with 148 farmers, a method that researchers commonly use to obtain data on pesticide use. However, surveys alone are not sufficient, as they do not provide researchers with an understanding of the all-important context—agricultural, climatic, political economic, cultural, etc.—in which farmers' pesticide-use decisions take place. Thus, rather than relying on a face-to-face survey alone, I adopted a multimethod political ecological approach resembling Burawoy's (2000) extended case methodology. The extended case method involves ethnographic participant observation, interviews, library research, and other techniques. It is extended in four ways, with the extension of (1) the observer into the world of the subject, (2) observations over space and time, (3) the micro situation to macro forces, and (4) findings to inform and modify social theory.

To this end, I complemented the survey with a number of other methods, including semistandardized interviews with a number of actors in important positions vis-à-vis agriculture (produce buyers, agricultural input salespeople, vegetable nursery owners, agronomists, and surveyed farmers); focus group discussions with export farmers as I was starting my research; participant observation; planting a test plot of comparative plantings of different national market and export vegetable varieties, which involved learning from farmers how to grow and sell vegetables; participatory pest mapping; and informal conversations with a number of people in the study site. My methods also included extensive library research in Costa Rica and mining of publicly available data sets from the U.N. Food and Agriculture Organization and the U.S. Food and Drug Administration.

These methods were integrated to help me create an understanding of the local context of pesticide use and vegetable production and their relationships to macro forces, or, as Blaikie and Brookfield (1987) emphasize, linking land users' decisions to the broader political economy. Thus, the study design integrates what Sayer (1992) terms "extensive research," which involves a large sample of one group (farmers), and "intensive research," which aims to understand the causal relationships between different groups (farmers, produce buyers, and agrochemical salespeople) and the structures that influence them through the operation of various mechanisms. More details are below.

Farmer Survey Conducted through Face-to-Face Interviews

My primary method for the 2003–2004 research involved a survey of 148 farmers I conducted through standardized, in-depth interviews[1] of vegetable farmers in Northern Cartago and the Ujarrás Valley (see Appendix 1 in Galt 2006 for the survey instrument). The vast majority of farmers in the area are the on-farm decision makers about pesticide applications, making them the appropriate decision makers for the survey. For the few farming firms included in the survey, in which management responsibilities were divided between personnel, the farm manager in charge of spraying decisions was surveyed.

The survey included questions regarding personal, household, resource, market, crop, farm, and field information. It also involved many detailed questions on various aspects of pesticide use, including specific pesticides used, pesticide dose, and preharvest intervals (PHIs), all asked in relation to specific crop varieties and field locations. Additionally, the last section of the survey, which was recorded with an audio recorder, allowed me to collect significant qualitative data concerning farmers' understandings of market requirements for export and the national market (including contract provisions and cosmetic and pesticide residue requirements), the effects of market requirements on their management decisions, organic agriculture and obstacles to it, and the susceptibility of various varieties to pests and diseases.

I conducted the survey between April 23, 2003, and January 4, 2004, largely corresponding to the rainy season known as *invierno* ("winter") in the study site. During the rainy season, rain falls on most days, and cloud cover can be very persistent. Since there are differences in pesticide use between the dry and rainy seasons, with fungicides applied more heavily during the rainy season and insecticides applied more heavily in the dry season (Galt 2001; Jiménez París 1996), in the survey I asked farmers about their pesticide use during the rainy season. Thus, for the purposes of the data in this book, pesticide-use data pertains to the rainy season.

Sampling

Since one major goal of my study was to compare production practices of farmers producing for the export and national markets, I had to create sampling strategies that would gather enough information on both groups of farmers. By necessity, sampling was different for these groups. Export farmers in Northern Cartago who produce minivegetables are few in number compared to national market farmers. For minivegetable export farmers, I started with the farmer contacts I had made in 2000. In my previous research I became acquainted with a number of farmers who grow for one minivegetable exporter, so I contacted them in 2003 to participate in the survey. I then used a snowball technique (Patton 2002, 237) of asking them for the contact information of other farmers oriented to the export and/or national markets. For minivegetable farmers who produce for the other minivegetable exporter, I contacted the manager of the export firm and asked for a list of farmers who sell their produce to the exporter. Instead of using this list directly, I used the contacts I had in the community of Cipreses to contact export farmers on the list, since I did not want to be seen by the export farmers as being affiliated with any exporter. I then used the snowball technique of asking these export farmers for the contact information of other farmers in their networks.

Surveying of national market farmers and chayote export farmers occurred in part through the snowball technique described above, since I obtained the contact information for some of them from export farmers. Another important way of finding these farmers was to approach them while they were working their fields near a number of towns in different locations. Additionally, I made important contacts, who also became good friends and key informants, in a number of towns and locations: Calle Naranjo and Birrís in the Ujarrás Valley and Cot, Cipreses, San Martín de Santa Rosa, and Buenos Aires de Pacayas in Northern Cartago (see figure 1.1). These key informants introduced me to many farmers in their communities and in surrounding towns. Sampling in this manner included farmers with different landholding sizes, since I told the key informants that I wanted to include a range from small-scale to large-scale farmers. Although it is impossible to tell if these sampling techniques sampled a subpopulation that is representative of farmers of the area since the last Costa Rican census of agriculture was produced in 1984 (Dirección General de Estadística y Censos 1986), I believe that the combination of the sampling strategies helped to avoid any serious bias concerning farm size or other important characteristics. For this reason, throughout the book I refer to farmers in the samples as simply "export farmers" and "national market farmers." I give sample sizes according to the number of farmers growing specific crops that I discuss.

Preharvest Interval

Pesticides are the result of instrumental knowledge applied in a reductionist way to overcome one particular problem in the agricultural production system. As such, of the many socioecological phenomena studied by political ecologists, pesticide use can be quantified relatively easily, since it involves the application of precisely measured amounts of substances to particular pieces of land. Few other land-use practices as objects of study are so neatly regimented in political ecology. Thus, quantitative methods can be applied more effectively to pesticide use than to many other objects of political ecology, such as social movements and environmental discourses.

Data collected by researchers concerning farmers' respect of the required time between spraying and harvest (the PHI) can be very different depending on how survey questions are asked. Valverde G., Carazo Rojas, and Araya Rojas (2001, 34–35) in their study of Costa Rican farmers' practices relating to pesticide residue levels on fresh produce report that "practically none of the farmers interviewed mentioned applying pesticides a week before harvest, nor in the post-harvest stage." It is, however, unlikely that farmers were reporting their actual practices because of the crops they grow and the climatic conditions. Many vegetables, such as sweet pepper, chayote, squash, and tomato, consistently produce fruit when they are in their harvest stage, with harvests every two to four days for several months. Additionally, the vegetable varieties available to Costa Rican farmers, with the exception of chayote, were generally created for temperate areas and lack adaptation to tropical climates, resulting in "very high pest damage" (Barbosa 2000, 105). Costa Rica's tropical climate is extremely conducive to pests and pathogens. Due to these circumstances, farmers in Costa Rica typically spray vegetable crops at least once per week. Even if a vegetable farmer sprays immediately after a harvest, the longest amount of time between the spray and the next harvest is three to four days, not a week or more as the farmers interviewed by Valverde G., Carazo Rojas, and Araya Rojas (2001) report.

This discrepancy exists because of the way survey questions are asked. During the course of my interviews I found that farmers rarely voluntarily say that they leave only a few days between spraying and harvesting if a researcher asks, as I did in my first few surveys, "How many days are there between the last spray and the harvest?" Farmers typically answered between one to four weeks, even for continuously harvested crops for which this would be impossible. In contrast, I found that if I asked in regard to these continuously harvested crops "When during the week do you harvest?" and the response was "Every Monday, Wednesday, and

Friday" and then I followed up with "So when do you spray?" the answer would be something like "Every Monday after harvest" or "On Saturdays." I interpret this difference in response to mean that farmers do not want to be seen as spraying very close to harvest, since this could be taken as a lack of consideration for the health of the consumer even though they find it a necessary practice to maintain their harvest and livelihoods. The data from the survey questions about PHI are used primarily in chapters 5 and 6.

Calculating Pesticide Dose and Intensities in Crop- and Spatially-Specific Ways

Obtaining complete data to calculate a crop's pesticide intensity in active ingredient (ai) per hectare (ha) per crop cycle involves a large set of questions. For this calculation, one needs to know each pesticide used on a crop and then five pieces of information about each pesticide: the amount of a spray mixture sprayed per unit of land, the amount of a pesticide-formulated product used in that spray mixture, the number of times that the mixture is sprayed per crop cycle, the percentage of active ingredient of that pesticide, and the actual size of the pesticide package used.

Questions about the first three pieces of information were organized by a large table in the survey while the last two pieces of information required finding the actual pesticide labels. The first question was "On crop ABC, which insecticides do you use?" I recorded this list, dividing it according to granulated and sprayed (liquid/powder) formulations. I then asked, "On crop ABC, which fungicides do you use?" I recorded these as well and then asked about herbicides used on crop ABC. With this list compiled, I then asked a series of questions about each pesticide used on crop ABC in order to calculate dose (in terms of kg ai/ha/cycle): (1) how many *estañones* (50 gallon drums) of spraying mixture they use per unit of land, (2) how much of a specific pesticide they use per *estañón,* and (3) the number of times they use the pesticide during the crop cycle. I proceeded in this manner for each crop in which I was interested (usually three).

Since most farmers grow a number of different vegetable crops and collecting production information on each vegetable would have been far too time intensive, I limited detailed data collection on production to three vegetables per farmer. My priorities for data collection were, in rank order, (1) the most important export crops that are also produced for the national market (squash, chayote, and green bean); (2) the most important national market crop in Northern Cartago (potato); (3) other crops grown for the open national market and, less commonly, the export and/or controlled national market (carrot, corn, and tomato); (4) other common national market vegetable crops in the area (broccoli, cabbage, cauliflower, and onion); and (5) uncommon vegetable crops in the area (cassava, fennel, jicama, spinach, etc.). Thus, data were gathered on many different vegetable crops, but some sample sizes are small. The definition of a crop used here is based on species and where applicable on subspecies/variety. For example, all potatoes (*Solanum tuberosum*) grown in the study site belong to the same species, and they are treated as one crop. I separate those that belong to the same species but are different subspecies or very different varieties, including cabbage (*Brassica oleracea* var. *capitata*), broccoli (*Brassica oleracea* var. *italica*), and cauliflower (*Brassica oleracea* var. *botrytis*); scallop squash (*Cucurbita pepo* var. *clypeata*) and zucchini (*Cucurbita pepo* var. *cylindrica*) (Lira Saade and Montes Hernández 1994); and dried beans and green beans (*Phaseolus vulgaris*, various varieties).

Between conducting farmer surveys, I visited agrochemical sales places throughout the area. These visits involved interviewing salespeople and examining pesticide packaging and labels to determine the percentage of active ingredient and the actual measured quantity of each specific pesticide container sold in the area (e.g., nine hundred grams in a package). This last piece of information is a correctional factor for farmers' tendency to round to the nearest whole unit (e.g., saying that a bag is a kilogram when a bag actually holds nine hundred grams).

Because of the considerable environmental variability in the study site, I collected all farmers' data in geographically specific ways. For all fields that a farmer had planted to the vegetables included in the survey, I asked about location, size, ownership, relative slope and fertility, irrigation systems, tillage method used, number of vegetable farms surrounding it, distance from the farmer's residence, the amount of fallow time between crops, and the number of years in consecutive vegetable production. I placed fields on the study site map by using place-names, roads, and relief features on the topographic maps. In most cases, the fields where the survey took place were the farmer's only field, making placement simple. I visited many other fields far from the survey location while spending time with farmers. For those fields I did not visit, I asked relative location compared to other farmers' fields that I knew, and I used this information—in addition to place-names, reported distance from the farmer's house, and topography—to place these fields. A handful of field locations (shown in figures 2.1 and 2.2) are therefore approximate.

Once I had collected information on field locations, I asked about differences in pesticide use when farmers reported growing the same crop in different fields. This included differences in greenhouses and plastic coverings versus open areas of planting, since many farmers pointed out that this makes a difference in pesticide use. If the farmer explained that there were differences in agrochemical treatments between the same crop grown in different fields, I entered data on the survey table as if these were separate crops, noted the fields to which the information pertained, and treated them as different field-specific spraying schedules during data analysis. If the farmer said that the same crops grown in different fields were treated the same, I noted the crop locations and varieties to which the pesticide information belonged and during data analysis also treated them as different field-specific spraying schedules.

Compiling this crop-specific and field-specific information on pesticide dose and frequency yielded what I call field-specific crop-spraying schedules for each farmer. With each farmer survey I collected information for between one and six field-specific crop-spraying schedules. A few of the interviews required an extra survey table, but the average number of field-specific crop-spraying schedules is 2.97 per farmer. The number and choice of field-specific crop-spraying schedules per survey depended on the number of crops and parcels the farmer had planted, my priorities in selecting specific crops (see above), and the amount of time the farmer was willing to spend being surveyed. The 145 farmer surveys yielded data on 430 field-specific crop-spraying schedules, of which 424 pertain to the category of vegetables. These data are used in the Introduction and chapter 2. Chapter 2 presents maps of pesticide intensity of field-specific crop-spraying schedules for export minisquash and national market potato production. These two crops were selected for the analysis because they are the crops for which I have a large sample size spread over a relatively large area of the study site.

Quantifying pesticide use and pesticide intensities of various crops can be used to move beyond vague statements common in the literature about the pesticide intensity of various crops. Once a national average of pesticide use per unit of land is estimated (see the Introduction), this can be used as the baseline against which to compare crops for which data is available or will be gathered. The system developed here and shown in figure I.1 is that crops under that national average are considered not pesticide intensive, while those from one to three times the national average are pesticide intensive, those from three to six times the national average are very pesticide intensive, and those six times or more than the national average are extremely pesticide intensive. While there is arbitrariness by necessity in drawing these lines, the distinctions are useful when one examines vegetable crops, since there is considerable variation within the category, ranging from not pesticide intensive (e.g., cassava) to extremely pesticide intensive (e.g., potato). Future comparative studies of pesticide use could make use of this system to move beyond vague statements that plague the literature on agroexport crops.

Household Asset Portfolios

Asset portfolio data were obtained through specific questions in the survey. Large resources—including land, house(s), pickup trucks, and tractors—are included as entries in the asset portfolio presented, as is relevant data concerning liquid capital: formal credit received, agricultural expenses in 2002, and agricultural profit in 2002. Rate of profit was calculated simply as profit divided by expenses. It is important to note that these expense and profit figures are from a farmer's total agricultural production, not specific crops. Since the vast majority of farmers plant a variety of vegetable crops, there is some potential interference from these other crops. Nevertheless, for both minisquash and potato farmers (the focus of chapter 2), these crops typically dominate farmers' agricultural production, so in most cases their expenses and profits strongly reflect the economic outcomes of their production of these specific crops.

Interviews and Participant Observation

I conducted interviews with twenty-three nonfarmer individuals involved in the agrifood system of the area, including five national-market produce buyers, seven export firm produce buyers and/or managers, one export firm's agronomist, six farm input retail salespeople (including one organic input provider), two agronomists not affiliated with produce buyers or input sales places, and two managers of vegetable nurseries. I used semistandardized protocols for people in different positions. I also conducted in-depth, open-ended interviews as follow-ups with twelve surveyed export farmers to further understand issues that arose during my fieldwork and in the survey. The interviews were transcribed and coded according to the themes that emerged during my fieldwork, including differences in pesticide use between markets, contract requirements, traceability systems, and specific pesticides such as methamidophos. Chapter 4 relies on the qualitative data from the exporter interviews, but insights from these interviews inform the entire book, as they together with the survey shaped my understanding of the agricultural and political economic context of the study site.

The interviews were supplemented by participant observation and a very large number of informal conversations with people in the area about agriculture, crops, pesticides, and markets. This included spending time with farmers on their farms, observing spraying practices, and participating in meetings of export farmer organizations. Visiting farms unannounced often resulted in witnessing pesticide mixing and application, which was used as a general check on the pesticide use that farmers reported during the survey.

Mapping of Climatic Conditions and Pest Pressure

Chapters 1 and 2 rely on cartographic representation of environmental data from secondary sources and pesticide-use data from the farmer survey. To merge environmental geography and farm- and household-level data, I used Adobe Illustrator with MAPublisher plug-ins to digitize Costa Rica's 1:50,000 topographic maps that pertained to the study site.

This included all of the Istarú quadrangle (Instituto Geográfico Nacional 1981a) and the northern quarter of the Tapantí quadrangle (Instituto Geográfico Nacional 1981b). The best available climatological data available for Northern Cartago and the Ujarrás Valley, from Barrantes F., Liao, and Rosales's (1985) climatological atlas of Costa Rica with data from 1961 to 1980, was added to the topographic maps. Unfortunately, the climatological data is less detailed than the base map.

In addition to representing environmental geographical data cartographically, I also engaged potato farmers in participatory mapping of pest intensity. I selected seven farmers in the survey who plant potato in a number of different fields in Northern Cartago. The method involved giving farmers a color print of the Istarú quadrangle on 8.5-by-11-inch paper and a blue and red pen. I oriented them to the map by pointing out well-known places and the location at which the mapping was taking place. I then asked farmers the question "Is the pressure of late blight on potato different in areas of Northern Cartago?" and following up with "Can you map the different zones?" Using the blue pen, farmers then drew zones and numbered them according to the strength of pressure from late blight for these zones. These questions and methods were then repeated by using the red pen for *polilla*, the local name for two species of tuber moths, *Scrobipalpopsis solanivora* and *Phthorimaea operculella*, that are important potato insect pests. These maps helped me understand the geography of pest pressure, as discussed in chapter 1.

Appendix 2   Detailed Pesticide Data

Figures A.1 and A.2 show detailed data about farmers' pesticide use on specific crops. Table A.1 provides the chemical class names specified as abbreviations in the figures. I constructed the figures based on Tufte's (1983, 1995) principles of data presentation, including allowing for micro and macro readings, and the inclusion of all data points in relation to each other and to relevant reference points. The figures show various aspects of farmers' pesticide use—chemical class, minimum preharvest interval (PHI), and dose—in relation to requirements on pesticide labels and regulation. The gray fill represents a problematic situation vis-à-vis label requirements in each aspect of pesticide use.

The dose section on the far right of the figures uses dots to plot doses used by each farmer in relation to the maximum recommended dose on the pesticide's label, which is shown as 100 percent. The number immediately to the left of the dose chart is the average dose. A gray fill here means that the dose used exceeds the label's requirements.

The PHI section, found left of the dose section, presents data on farmers' respect of PHI. The numbers are the result of subtracting the maximum required PHI on the label from farmers' minimum reported PHIs (i.e., the minimum number of days a farmer leaves between spraying and harvesting). Thus, a zero means that the farmer respects the PHI exactly, while a negative number means that PHI is violated—in other words, the farmer harvests too early, without enough time having passed for the pesticide to break down to relatively low levels. "Average minimum PHI" shows the average of the minimum PHIs reported by farmers who use that pesticide. The gray background means that it violates the PHI required by the label.

Regulatory information is presented to the left of PHI information. The rightmost checkboxes in the section show whether a particular active

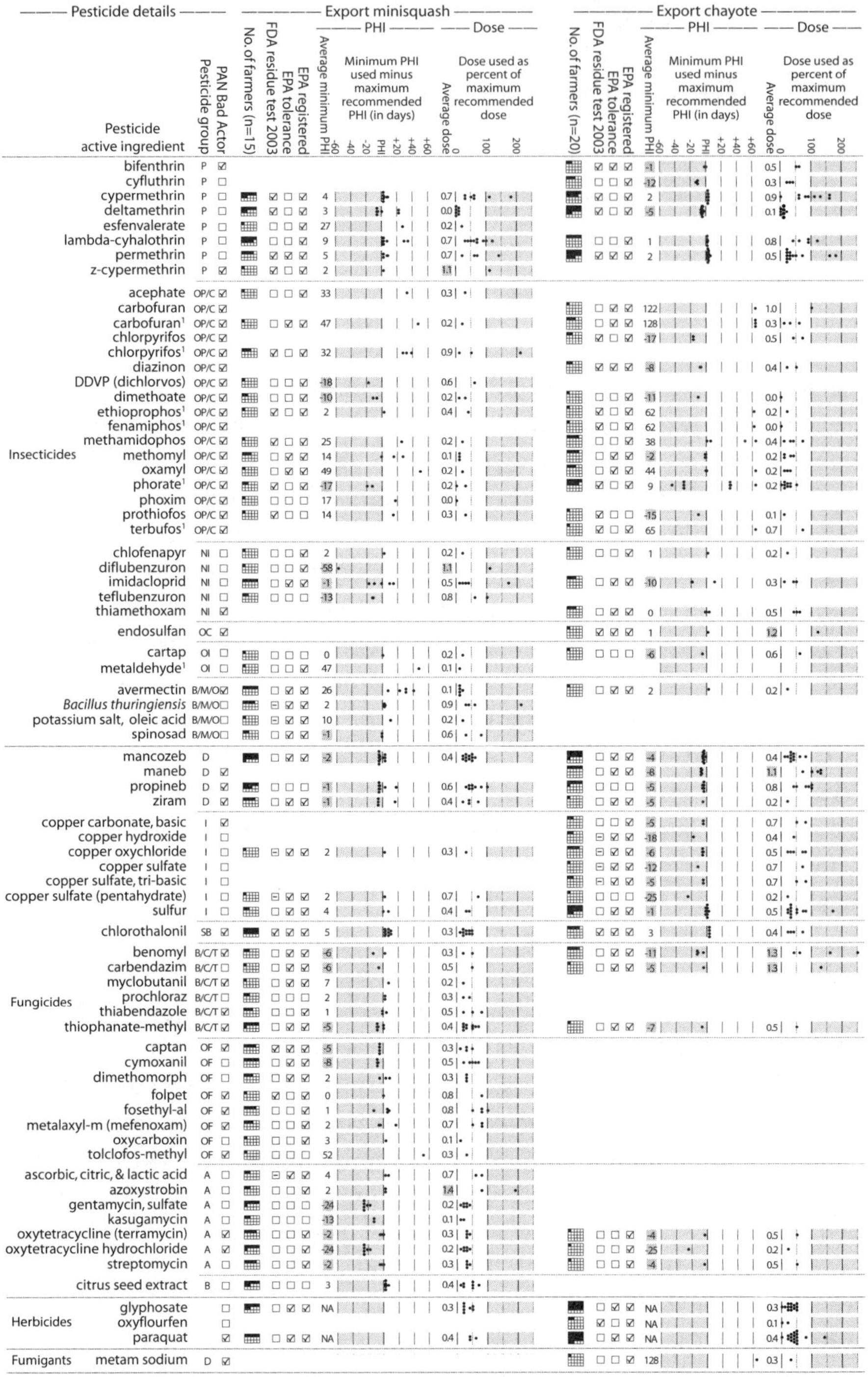

Figure A.1 Pesticide use on export minisquash and chayote in relation to U.S. regulation, Northern Cartago and the Ujarrás Valley, 2003–2004

[1] Applied as granulated formulations.
Sources: Author's farmer surveys 2003-04 & Costa Rican pesticide labels; EPA (2004) & FDA (2008).

	Pesticide active ingredient	Pesticide group	Export minisquash Avg minus max PHI	Export minisquash Avg dose %	Export chayote Avg minus max PHI	Export chayote Avg dose %
	bifenthrin	P	-1	0.5		
	cyfluthrin	P	-12	0.3		
	cypermethrin	P	4	0.7	2	0.9
	deltamethrin	P	3	0.0	-5	0.1
	esfenvalerate	P	27	0.2		
	lambda-cyhalothrin	P	9	0.7	1	0.8
	permethrin	P	5	0.7		
	z-cypermethrin	P	2	1.1	2	0.5
	acephate	OP/C	33	0.3		
	carbofuran	OP/C			122	1.0
	carbofuran[1]	OP/C	47	0.2	128	0.3
	chlorpyrifos	OP/C			-17	0.5
	chlorpyrifos[1]	OP/C	32	0.9		
	diazinon	OP/C			-8	0.4
	DDVP (dichlorvos)	OP/C	-18	0.6	-11	0.0
	dimethoate	OP/C	-10	0.2		
	ethioprophos[1]	OP/C	2	0.4	62	0.2
	fenamiphos[1]	OP/C			62	0.0
	methamidophos	OP/C	25	0.2	38	0.4
	methomyl	OP/C	14	0.2	-2	0.2
	oxamyl	OP/C	49	0.2	44	0.2
	phorate[1]	OP/C	-17	0.2	9	0.2
	phoxim	OP/C	17	0.0		
	prothiofos	OP/C	14	0.3	-15	0.1
	terbufos[1]	OP/C			-65	0.7
	chlofenapyr	NI	2	0.2	1	0.2
	diflubenzuron	NI	-58	1.1		
	imidacloprid	NI	-1	0.5	-10	0.3
	teflubenzuron	NI	-13	0.8		
	thiamethoxam	NI			0	0.5
	endosulfan	OC			1	1.2
	cartap	OI	0	0.2	-6	0.6
	metaldehyde[1]	OI	47	0.1		
	avermectin	B/M/O	26	0.1	2	0.2
	Bacillus thuringiensis	B/M/O	2	0.9		
	potassium salt, oleic acid	B/M/O	10	0.2		
	spinosad	B/M/O	-1	0.6		
	mancozeb	D	-2	0.4	-4	0.4
	maneb	D			-8	1.1
	propineb	D	-1	0.6	-5	0.8
	ziram	D	-1	0.4	-5	0.2
	copper carbonate, basic	I			-5	0.7
	copper hydroxide	I			-18	0.4
	copper oxychloride	I	2	0.3	-6	0.5
	copper sulfate	I			-12	0.7
	copper sulfate, tri-basic	I			-6	0.7
	copper sulfate (pentahydrate)	I	2	0.7	-25	0.2
	sulfur	I	4	0.4	-6	0.5
	chlorothalonil	SB	5	0.3	3	0.4
	benomyl	B/C/T	-6	0.3	-11	1.3
	carbendazim	B/C/T	-6	0.5	-5	1.3
	myclobutanil	B/C/T	7	0.2		
	prochloraz	B/C/T	2	0.3		
	thiabendazole	B/C/T	1	0.5		
	thiophanate-methyl	B/C/T	-5	0.4	-7	0.5
	captan	OF	-5	0.3		
	cymoxanil	OF	-8	0.5		
	dimethomorph	OF	2	0.3		
	folpet	OF	0	0.8		
	fosethyl-al	OF	1	0.8		
	metalaxyl-m (mefenoxam)	OF	2	0.7		
	oxycarboxin	OF	3			
	tolclofos-methyl	OF	52	0.3		
	ascorbic, citric, & lactic acid	A	4	0.7		
	azoxystrobin	A	2	1.4		
	gentamycin, sulfate	A	-24	0.2		
	kasugamycin	A	-13	0.1		
	oxytetracycline (terramycin)	A	-2	0.3	-4	0.5
	oxytetracycline hydrochloride	A	-24	0.2	-25	0.2
	streptomycin	A	-2	0.3	-4	0.5
	citrus seed extract	B	3	0.4		
	glyphosate	(Herbicides)	NA	0.3	NA	0.3
	oxyflourfen				NA	0.1
	paraquat		NA	0.4	NA	0.4
	metam sodium (Fumigants)	D			128	0.3

Figure A.2 — data table

	Pesticide details			Open national market			Export market		
	Pesticide active ingredient	Pesticide group	PAN Bad Actor	No. of farmers (n=26) / Costa Rican tolerance / Sanidad Vegetal test	Avg min PHI — Min PHI used minus min recommended PHI (days)	Avg dose — Dose used as % of max recommended dose	No. of farmers (n=15) / FDA residue test 2003 / EPA registered / EPA tolerance	Avg min PHI — Min PHI used minus min recommended PHI (days)	Avg dose — Dose used as % of max recommended dose
Insecticides	cyfluthrin	P			-11	0.2			0.7
	cypermethrin	P			4	1			0.0
	deltamethrin	P			5	0.0			3 / 0.0
	esfenvalerate	P							27 / 0.2
	lambda-cyhalothrin	P			13	[99] 0.8			9 / 0.7
	permethrin	P			2	0.5 [3]			5 / 0.7
	z-cypermethrin	P	☑						2 / 1.1
	acephate	OP/C	☑		-2	0.5			33 / 0.3
	carbofuran[1]	OP/C	☑		92	[92] 0.2			47 / 0.2
	chlorpyrifos	OP/C	☑		-17	0.2			
	chlorpyrifos[1]	OP/C	☑		12	1.2			32 / 0.9
	diazinon	OP/C	☑		-7	0.3			
	DDVP (dichlorvos)	OP/C	☑		-18	0.5			-18 / 0.6
	dimethoate	OP/C	☑		-12	1.3			-10 / 0.2
	ethion	OP/C	☑		-19	0.7			
	ethioprophos[1]	OP/C	☑						2 / 0.4
	fenamiphos[1]	OP/C	☑		62	[92] 0.1			
	malathion	OP/C	☑		0	0.5			
	methamidophos	OP/C	☑		-6	0.6			25 / 0.2
	methomyl	OP/C	☑			0.2			14 / 0.1
	oxamyl	OP/C	☑		39	0.2			49 / 0.2
	phorate[1]	OP/C	☑		3	0.4			-17 / 0.2
	phoxim	OP/C	☑						17 / 0.0
	prothiofos	OP/C	☑						14 / 0.3
	terbufos[1]	OP/C	☑		2	0.4			
	chlofenapyr	NI							2 / 0.2
	diflubenzuron	NI							-58 / 1.1
	imidacloprid	NI			-16	0.9			-1 / 0.5
	teflubenzuron	NI							-13 / 0.8
	thiamethoxam	NI	☑		-1	0.7			
	endosulfan	OC	☑		0	0.2			
	cartap	OI			-5	0.2			0 / 0.2
	metaldehyde[1]	OI							47 / 0.1
	thiocyclam	OI			-12	0.4			
	avermectin	B/M/O	☑						26 / 0.1
	Bacillus thuringiensis	B/M/O							2 / 0.9
	potassium salt, oleic acid	B/M/O							10 /
	spinosad	B/M/O							-1 / 0.6
Fungicides	mancozeb	D	☑		-5	0.4			0.4
	maneb	D	☑		-8	0.8			
	propineb	D	☑		-5	0.5			-1 / 0.6
	ziram	D	☑		-5	0.3			-1 / 0.4
	copper carbonate, basic	I	☑		-5	1.3			
	copper hydroxide	I			-18	0.8			
	copper oxychloride	I			-5	0.4			2 / 0.3
	copper sulfate	I			-12	0.8			
	copper sulfate, tri-basic	I			-5	1			
	copper sulfate (pentahydrate)	I							
	sulfur	I				0.3			4 / 0.4
	chlorothalonil	SB	☑		2	0.3			5 / 0.3
	benomyl	B/C/T	☑		-13	1.7 [44]			-6 / 0.3
	carbendazim	B/C/T	☑		-5	1			-6 / 0.5
	myclobutanil	B/C/T	☑		-1	2.9 [29]			7 / 0.3
	prochloraz	B/C/T			2	0.2			2 / 0.3
	thiabendazole	B/C/T	☑						1 / 0.5
	thiophanate-methyl	B/C/T	☑		-7	0.2			-5 / 0.4
	captan	OF	☑		-5	0.3			-8 / 0.5
	cymoxanil	OF			-8	0.3			-8 / 0.5
	dimethomorph	OF							2 / 0.3
	flutolanil	OF			-10	0.5			
	folpet	OF	☑						0 / 0.8
	fosethyl-al	OF	☑						1 / 0.8
	metalaxyl-m (mefenoxam)	OF							2 / 0.7
	oxycarboxin	OF							3 / 0.1
	tolclofos-methyl	OF	☑						52 / 0.3
	ascorbic, citric, & lactic acid	A							4 / 0.7
	azoxystrobin	A							2 / 1.4
	gentamycin, sulfate	A							-24 / 0.2
	kasugamycin	A							-13 / 0.1
	oxytetracycline (terramycin)	A	☑						2 / 0.3
	oxytetracycline hydrochloride	A	☑						-24 / 0.2
	streptomycin	A							2 / 0.3
	citrus seed extract	B			3	0.2			3 / 0.4
Herbicides	glyphosate				NA	0.3			NA / 0.3
	linuron		☑		NA	0.2			NA
	paraquat		☑		NA	0.3			NA / 0.4

[1] Applied as granulated formulations.

Sources: Author's farmer surveys 2003–04 & Costa Rican pesticide labels; La Gaceta (1997) & Rodríguez Solano (1994); EPA (2004) & FDA (2008).

Figure A.2 Pesticide use on open national market and export squash in relation to regulation, Northern Cartago and the Ujarrás Valley, 2003–2004

Table A.1 Chemical classes and modes of action for 122 pesticide active ingredients used in Northern Cartago and the Ujarrás Valley

Abbreviation	Chemical Class	Active Ingredients[a]	Mode of Action[b]
A	azole	{ bromuconazole, cyproconazole, imazalil, myclobutanil, prochloraz, propiconazole, tebuconazole, triadimenol	F
AB	antibiotic	kasugamycin	B, F
Al	aldehyde	metaldehyde	I
An	anilide	flutolanil, oxadixyl	F
APA	aryloxyphenoxy propionic acid	fluazifop-p-butyl, haloxyfop	H
B	benzimidazole	benomyl, carbendazim, thiabendazole, thiophanate-methyl	F
BiP	bipyridylium	paraquat	H
Bot	botanical	avermectin (I), citrus seed extract (B, F)	Misc
BU	benzoylurea	diflubenzuron, flufenoxuron, lufenuron, novaluron, teflubenzuron	I
C	carbamate	benfuracarb, carbofuran, carbosulfan, methomyl, oxamyl	I, N
CA	chloroacetanilide	alachlor	H
CD	cyclohexenone derivative	clethodim	H
CH	carbohydrate	validamycin	F
CM	carboxamide	carboxin, oxycarboxin	F
C-N	chloro-nicotinyl	acetamiprid, imidacloprid	I
DC	dithiocarbamate	mancozeb, maneb, metam-sodium (Fum), metiram, propineb, zineb, ziram	F

DE	diphenyl ether	oxyfluorfen	H
DM	dicarboximide	iprodione	F
I	imidazole	fenamidone	F
In	inorganic	copper carbonate basic, copper hydroxide, copper oxchloride, copper sulfate, copper sulfate pentahydrate, copper sulfate tribasic, sulfur	F
M	morpholine	dimethomorph, tridemorph	F
Mic	microbial	*Bacillus thuringiensis*, spinosad	I
N	nereistoxin	cartap, thiocyclam hydrogen oxalate	I
Neo	neonicotinoid	thiamethoxam	I
OtC	other carbamate	propamocarb	F
OC	organochlorine	endosulfan (I), PCNB (F, N)	Misc
OP	organophosphate	acephate, chlorpyrifos, DDVP, diazinon, dimethoate, ethion, ethioprophos (I, N), fenamiphos (I, N), malathion, methamidophos, methyl parathion, naled, phorate, phoxim, prothiofos, terbufos, tolclofos-methyl (F)	I
OT	organotin	fentin hydroxide, triphenyltin acetate	F
Ox	oxadiazine	indoxacarb	I
Oxa	oxazolidinedione	famoxadone	F
P	pyrethroid	bifenthrin, cyfluthrin, cypermethrin, deltamethrin, esfenvalerate, ethofenprox, lambda-cyhalothrin, permethrin, z-cypermethrin	I
Py	pyrazole	chlorfenapyr, fipronil	I
Ph	phosphonoglycin	glyphosate	H

continued

Table A.1 Continued

Abbreviation	Chemical Class	Active Ingredients[a]	Mode of Action[b]
PS	phenylsulfamide	dichlofluanid	H
S	sulfonylurea	halosulfuron-methyl	F
SB	substituted benzene	chlorothalonil	F
So	soap	oleic acid potassium salt	I
St	strobin	azoxystrobin	F
T	triazine	atrazine, cyromazine (I), terbuyryn	H
Th	thiophthalimide	captan, folpet	F
Tr	triazinone	metribuzin	H
U	urea	linuron (H), pencycuron (F)	Misc
Un	unclassified	ascorbic acid, bentazon, citric acid, cymoxanil, ethofenprox, fosetyl-al, gentamycin sulfate, lactic acid, oxytetracycline, oxytetracycline hydrochloride, streptomycin	Misc
VD	valine diamide	iprovalicarb	F
X	xylylalanine	metalaxyl, metalaxyl-m	F

[a] Modes of action that differ from the normal mode(s) of action for the class are in parenthesis after specific active ingredients.
[b] Modes of action are: B=bactericide/microbiocide, F=fungicide, Fum=soil fumigant, H=herbicide, I=insecticide, N=nematicide.

ingredient is registered by the U.S. Environmental Protection Agency (EPA) for any agricultural use. The middle checkboxes show whether the EPA allows residues of a particular pesticide to exist on the crop in question (see figure A.1). A check signifies the existence of an EPA tolerance, and a dash shows an exemption, both of which mean that the EPA permits that specific pesticide-crop combination. The absence of a check means that the pesticide-crop combination is illegal in the United States.

To the left of the regulatory section is data on the number of farmers in the sample who use the pesticide on the crop in question. Filled-in boxes represent the number of farmers, while the total number of boxes show the sample size.

Figure A.2 uses a very similar design but compares across market orientation by showing squash produced for the open national market and the export market. The comparison is aided by signaling whether the pesticide is classified as a bad actor by the Pesticide Action Network (Orme and Kegley 2004).

In addition to its comparative nature, figure A.2's main difference in relation to figure A.1 is that the regulatory information is specific for each market. The export regulations for figure A.2 are the same as in figure A.1, but the national market regulations refer to Costa Rican tolerances and residue tests. "Costa Rican tolerance" refers to the existence of a legal tolerance for the pesticide on the specific crop, as established in La Gaceta (1997). "Residue test" refers to whether the pesticide would be detected by Costa Rica's pesticide residue testing program (Rodríguez Solano 1994).

Notes

Introduction

Small portions of this chapter, including figure I.1, are reprinted from Ryan E. Galt, "Pesticides in Export and Domestic Agriculture: Reconsidering Market Orientation and Pesticide Use in Costa Rica," *Geoforum* 39, no. 3 (2008): 1378–92. Reprinted with permission of Elsevier.

1. All interviewees' names have been changed to maintain confidentiality.

2. All translations from Spanish to English are mine. I use italics for Spanish words not commonly used in English.

3. There is a glossary of pesticide technical terms at the end of the book. Many of the larger arguments that I make require an understanding of these technical details, so I have included those pieces that I feel are needed for those less familiar with these aspects of pesticides.

4. Other estimates are similar. In 1999, Costa Rica's pesticide use per cultivated hectare averaged 23.2 kg/ha/year, higher than all other Central American countries and double the regional average of 11.5 kg/ha/year (Arbeláez and Henao H. 2002, 10–11). In recent years, Costa Rica's pesticide use appears to have increased even more, to 51.2 kg/ha/year, with Colombia at a distant second of 16.7 kg/ha, thereby maintaining its world leadership in pesticide intensity (World Resources Institute, cited in Andréu 2011).

5. Pesticides consist of an active ingredient, which acts against the pest organism, and other ingredients (often incorrectly called inert ingredients) that help the active ingredient or its application. Together these make up the formulated product. In quantifying pesticide use, active ingredient is more commonly used than formulated product, a convention I follow here.

6. I use three market segments: the open national market, the controlled national market, and the export market (these are explained in depth in chapter 6). Based on Chaverri's (1999) estimate of average agricultural pesticide intensity in Costa Rica, I use a classification system of pesticide intensity for the different vegetables by standardizing all measures of pesticide intensity by kilograms of active ingredient per hectare per week (kg ai/ha/week). This system has the following categories relative to the national average: "not pesticide intensive" is less than the Costa Rican average,

"pesticide intensive" is one to three times the Costa Rican average, "very pesticide intensive" is three to six times the Costa Rican average, and "extremely pesticide intensive" is more than six times the Costa Rican average. While these divisions are necessarily arbitrary, the benefit of this classification system is that it allows for more precise gradations of the term "pesticide intensive," a phrase that is too often used without supporting or comparative data.

7. This pertains to crops with a sample size greater than four. One beet farmer sprays his beets very intensively, making the average estimate very high and likely unrepresentative of beet production in the area.

8. This literature spans many disciplines and fields: geography (Bee 2000; Brohman 1996; Gwynne 1993, 1999; Gwynne and Ortiz 1997; Murray 1997, 1998; Thrupp 1996; Thrupp, Bergeron, and Waters 1995), sociology (Collins 1995; Collins and Krippner 1999; Mannon 1998, 2005; Murray 1991, 1994; Murray and Hoppin 1992; Raynolds 1994, 1997, 2000; Stonich 1992, 1993, 1995; Stonich, Murray, and Rosset 1994), anthropology (Andreatta 1997, 1998a, 1998b; Fischer and Benson 2006; González 2001; Hamilton, Asturias de Barrios, and Tevalán 2001; Hamilton and Fischer 2003, 2005), agricultural economics (Barham, Carter, and Sigelko 1995; Barham et al. 1992; Carletto, de Janvry, and Sadoulet 1999; Carter and Barham 1996; Conroy, Murray, and Rosset 1996; Sullivan et al. 1999), and environmental studies (Wright 1986, 1990).

9. These areas include but are certainly not limited to Cameroon (Gockowski and Ndoumbe 2004), Nepal (Pujara and Khanal 2002) and the broader Hindu Kush–Himalaya (Tulachan 2001), Malaysia (Midmore et al. 1996), the Philippines (Lewis 1989, 1992), Mexico (Sánchez Saldaña and Betanzos Ocampo 2006), Guatemala and Ecuador (Horst 1987; Keese 1998), Colombia (Bojacá et al. 2013), and Peru (Zimmerer 1991, 1999).

10. There is a large amount of literature on these impacts (Buhs 2002; Castillo, de la Cruz, and Ruepert 1997; Colborn, vom Saal, and Soto 1993; Edwards 1993; Guan-Soon and Seng-Hock 1987; Lehman 1993; Pimentel 1995; Pimentel et al. 1992; Pimentel and Greiner 1997; Sarmah, Müller, and Ahmad 2004; von Düszeln 1991; Wilson 2000).

11. This literature grows out of earlier work, especially that of Friedland (1984) who called his approach "commodity systems analysis." He focused on iceberg lettuce and other California-based commodity chains (grapes and processing tomatoes) and set the groundwork for analysis of commodity chains by focusing on production practices, the organization of the farm unit and grower organizations, labor as a factor of production, the production of science and its application, and marketing and distribution networks. Later, commodity chain analysis sought to describe the workings of various actors along the places through which commodities pass (Gereffi and Korzeniewicz 1994). Commodity chain studies are typically grounded at least partially in a Marxian understanding of commodities, including a critique of their fetishization, that is, seeing commodities as adequately represented by their prices rather than as a

set of hidden socioecological relations that produce them. In the 1990s, edited volumes with empirical analyses and theoretical framings made the analysis of agrifood commodity chains a major part of the political economy of agriculture (Goodman and Watts 1997; McMichael 1994, 1995).

12. One of the diseases featured prominently in the book, late blight (*Phytophthora infestans*), an oomycete, was originally part of the Fungi kingdom but is now classified in the kingdom Chromista. Oomycetes are referred to as being fungus-like because they share certain fungal characteristics. To simplify the story and word use, I refer to oomycetes as fungi even though this is not technically correct according to current taxonomy.

Chapter 1

1. On one occasion I witnessed a young boy helping direct a farmer's application by acting as a moving marker.

2. Costa Rican agricultural census data exist for chayote and potato but not other vegetable crops, so these are my estimates.

3. Several other diseases and insects are important constraints on production. Other fungal pathogens for potatoes are early blight, *Alternaria solani* (Ell. and G. Martin) Sor., and *Rhizoctonia solani* Kühn. Farmers control early blight with the same fungicides used for late blight control. *R. solani* resides in the soil and can be spread by tubers (Jackson 1983) and is controlled with soil sterilants such as PCNB. Two important bacterial diseases of potato in Costa Rica are bacterial wilt, *Pseudomonas solanacearum* E. F. Smith, and blackleg, *Erwinia carotovora* var. *atroseptica* (van Hall) Dye. Bacterial wilt exists below twenty-two hundred masl in the study site and stays in the soil for many years (Jackson 1983, 104). It can be reduced by rotations, especially those with pasture grasses.

4. I view off-farm capital as different from on-farm capital, which refers to the social units directly engaged in production and their mobilization of capital and resources in this pursuit. Although certainly not all farmers are fully capitalist in rationality (Mooney 1988), all but the most subsistence-oriented farmers engage in commodity markets for both inputs and outputs, thereby making their actions a part of broader circuits of capital accumulation, even if accumulation is not their primary goal. I therefore use on-farm capital to discuss the farming unit, which is usually households in the study area, and the capital it controls.

5. I spent considerable time in agrochemical retail stores. These are small buildings with a retail floor and back storeroom. Solid walls of the vast array of bottles and bags of pesticides offered sit on the shelves behind the service counter, joined by foliar nutrients—crop nutrients prepared in a spraying mixture and sprayed on crop foliage—that farmers also commonly use. Large sacks of fertilizer and other miscellaneous farm equipment are easily accessible on the retail floor, including rubber boots, rubberized clothing for spraying, and small implements such as machetes and shovels.

In addition to their function of selling farm inputs (for the benefit of off-farm capital), these retailers serve as places where farmers commonly socialize.

6. Most research has concentrated on the diamondback moth (*Plutella xylostella*), the primary insect pest of cabbage and other cole crops. By the 1990s, many studies in the study site showed that the diamondback moth had developed resistance to three synthetic pyrethroids (Blanco, Shannon, and Saunders 1990); deltamethrin and methamidophos (Monge V. et al. 1996); and *Bacillus thuringiensis* (Bt), a bacterial insecticide (Cartín L. et al. 1999; Monge V. et al. 1996; Perez and Shelton 1997). Blanco, Shannon, and Sanders (1990) found that between moth populations in Pacayas, Zarcero, and Santa Cruz de Turrialba, those in Pacayas—the only site in Northern Cartago—were the most resistant to deltamethrin. The population in Pacayas was also most resistant to lambda-cyhalothrin before its widespread use, suggesting that cross-resistance to new synthetic pyrethroids already existed in 1990. Only one study examines plant pathogen resistance (Páez et al. 2001). Late blight was examined for resistance to the fungicide metalaxyl in Northern Cartago and Zarcero. In Northern Cartago, 52 percent of late blight samples collected between July 1999 and January 2000 were resistant to metalaxyl (Páez et al. 2001, 37). There is likely considerably more fungicide resistance developed by late blight and other fungi in the area.

Chapter 2

1. The distinction between time-wage work and piece-wage work is crucial here: "The capitalist contractor, in dealing with his time-wage employees, faces serious constraints on his efforts to increase the production of surplus value. He faces few such constraints in dealing with his piece-wage employees—'independent' family farmers under contract, a propertied labor force that is non-unionized, self-directed, and willing to work without the guarantees of minimum wage, job security, insurance, and other benefits commonly demanded by time-wage employees" (Davis 1980, 143).

2. The geography of pesticide use, especially spatial differences at a landscape scale, is an underdeveloped area of knowledge. Some epidemiological studies about the connections between pesticide use and disease have examined spatial aspects of pesticide use (Garry et al. 1996), although these are not interested in broader people-environment interactions or social processes. Only a handful of geographical studies exist on the spatial aspects of pesticide use (Kansakar, Khanal, and Ghimire 2002; Li et al. 2004), but these are not attentive to political economic processes. Other studies by geographers address pesticide use and farmer decision making but fail to place it within a detailed environmental and geographical context at the landscape scale (e.g., Grossman 1998) or employ cartographic methods but not a detailed spatial understanding of pesticide use (e.g., Pujara and Khanal 2002). Combining a political ecological understanding with the spatial aspects of pesticide use might yield interesting and insightful results in other areas as well.

3. His pesticide intensity data is not included on the map, since he did not have export minisquash planted there at the time of the survey.

4. The lower temperatures of the higher elevations also means that bacterial wilt (*Pseudomonas solanacearum*), known locally as *maya* and endemic to many parts of the lowland tropics and some parts of montane Latin America (Horton 1987), is not a limit on production, since it is not problematic above 2,500 masl in Northern Cartago, although it is "particularly dangerous" at elevations below 1,500 masl (Sáenz Maroto 1955, 27).

5. Morales and Perfecto (2000) report a similar finding for corn in the Guatemalan highlands versus the lowlands.

6. I do not have the data to do a full analysis of the origins of the excess profits of farmers employing a spatial strategy, but in my conversations with farmers they saw it as the difference between payments to land (which are about equal in and out of the cloud belt) and differences in input use (with substantial savings in agrochemicals and higher quality of production existing outside of the cloud belt). David Ricardo (1971) was the classical political economist to first provide a theory of rent, which included differential rent from fertility differences. He argued that land rents are based on payments for the productive powers of land, which means that owners of more productive land can charge rents once land of lower productive powers is brought into production. One problem is that Ricardo's ecological conceptions were very thin, as he defined rent as "that portion of the produce of the earth, which is paid to the landlord for the use of the original and indestructible powers of the soil" (91). Marx critiqued Ricardo's concept of rent on a number of levels, including by arguing that differential rent can come from "*progressive improvement*" of agriculture (Marx 1968, 241, emphasis in original, cited in Barnes 1984, 128). Research on the importance of soil amendments and agricultural practices in affecting soil properties and productivity has borne this out, as has work showing the real possibilities that productive powers can be destroyed through erosion, salinization, acidification, etc. (see also my discussion of capitalism's second contradiction in chapter 1). There is also a question of farming system complexity when it comes to setting land rents: in agricultural systems with one or a few crops (upon which Ricardo's theory is based), the differences in land that cause differentials in production can be clearly understood, but in regions where dozens of crops can be grown, it is not clear how specific characteristics of land will impact each crop, since these characteristics differentially impact different crops. The problems of Ricardo's conception extends into input use as well. "To examine agricultural rent within a pure labor theory of value, Ricardo assumed that no intermediate goods are used in production" (Barnes 1984, 128). While this might have worked in conceptualizing English agriculture at the time, it certainly does not apply to Costa Rican vegetable farming. Class power of landlords and the use of sharecropping to access land, as with some of the larger minisquash growers, might also play an interesting role that I have not investigated. These are all possibilities for future analysis.

Chapter 3

1. In terms of Costa Rican chayote consumption, Trejos (1966, 114) mentions that chayote root is common and that the fruit of chayote "is served on all tables."

2. It is very likely that the "Other" category in the census contains some farms dedicated to vegetable production. The question in the census asked the farmer to identify the type of farm, which is fairly simple if it is dedicated to one crop (e.g. a banana farm is a *bananera*). For a mixed vegetable farm, it is quite possible that farmers identified it is a farm dedicated to a certain vegetable rather than a "horticultural" farm that the census categories reflect.

3. Chiverre, of the species *Cucurbita ficifolia* (Bolaños Herrera 2001), is now grown by only a few farmers in the area and hardly ever commercially. Sauer (1969) suggests that it originated in highland Mexico and Central America, although earlier researchers considered it an Old World crop. Harvested when the size of a watermelon, it is cooked as a dessert in the area, with its spaghetti-like strands mixed with brown sugar.

4. In 1967 the Cartago Beef Packing Company began exporting beef to the United States from Cartago. Documentation does not provide any information on whether old dairy cows became part of beef exports, although it does note that transportation costs for cattle were higher than for beef-packing companies on the Pacific coast, where most Costa Rican cattle ranching occurs (Vianna and Wierer 1971). The company is no longer in business.

5. Phytosanitary controls are those aimed at eliminating the introduction of pest organisms into an area. These controls typically involve inspections, quarantines, and/or the use of pest-free certification systems for imported plant materials, especially seeds.

6. The replacement of inorganic copper compounds with synthetic fungicides was likely the result of their superior performance, since copper products reduce the foliar area and "harden" the plant, leading to lower harvests (Montaldo 1984, 446). Copper sprays can be used in rainy regions and when the plants are already developed (Montaldo 1984), but aside from this last use, they are now little used in potato production by farmers in the area.

7. The fact that potatoes are produced on about thirty-three hundred hectares per year, which is 0.76 percent of cultivated land in Costa Rica (Secretaría Ejecutiva de Planificación Sectorial Agropecuaria 2004, 51), shows the very intensive nature of the production system.

8. Because of the strong competition between the two minivegetable exporters, they disagree about who first introduced minivegetables into Northern Cartago. Based on the two stories I was told, they were either first brought from the United States by the manager of one of the firms or by the other firm's large-scale contract farmers who traveled to the United States looking to diversify production due to unstable and declining national market vegetable prices.

9. Although I point to problems of national provisioning, my research should not be mistakenly interpreted to mean that export production is the way forward for

Northern Cartago and the Ujarrás Valley nor for Latin America. Relying only on export markets is risky since this depends on the economic well-being of the export market, a lesson that Latin America learned well when the Great Depression hit the United States. Development based on fulfilling local needs, especially articulated development (de Janvry 1985), is essential to the Latin American population and the world generally, and agroexport promotion by itself will not accomplish this. My larger point is that agroexport expansion is nondeterministic in terms of development and environment: it can lead to growth that reduces or exacerbates inequality (Carter, Barham, and Mesbah 1995) and may result in substantial environmental degradation or pesticide rationalization, as is the case in Northern Cartago and the Ujarrás Valley.

10. Since CAFTA-DR has gone into effect, U.S. potato imports into Costa Rica have risen from $110,000 in 2005 to $930,000 in 2009, which is an increase from 4 percent to 12 percent of market share (Foreign Agricultural Service 2010, 3). A detailed investigation is needed to determine the impacts of these and future imports on Costa Rican potato farmers and their production practices, since the possibility of undermining potato farmers' livelihoods is very real.

Chapter 4

1. For reasons of confidentiality I do not use the actual names of the export firms of their employees, nor do I provide identifying details of the firms.

2. Whether these logbooks would be effective in determining actual malevolent actions is extremely doubtful, since one would likely not note an intentionally malevolent act. They may prove useful in determining the source of accidental biological or chemical contamination, but this does not appear to be the case given the difficulty that the FDA has had in tracking down salmonella and other food-borne pathogens at the farm level in recent years.

3. A General Accounting Office (Government Accountability Office since 2004) report extensively documents the common occurrence that produce shipments in violation of the EPA pesticide residue tolerances reach consumers and that importers responsible for this are rarely punished (General Accounting Office 1992).

4. Exporters explained it in a way similar to Freidberg's (2009, 139) account of early California fruit shipments: "growers' distance from their markets left them dependent on wholesale commission agents to find buyers. These merchants only kept the records that suited them. They could easily lie about the condition of the fruit on arrival, or how much it sold for, or whether it sold at all."

5. The reporter does not accurately identify the agency responsible, an understandable mistake given the fact that the USDA, the FDA, and the EPA are all involved in regulation of residues.

6. Cucurbits are known to absorb another organochlorine pesticide, dieldrin, through the roots, which has provoked residue problems in Mexican cantaloupe (Groth, Benbrook, and Lutz 2000).

7. This includes sixteen members of the farmer organization who are also partners in the business and thirty "producers," who are typically much smaller-scale farmers. Farmers and employees associated with Mini-Horta dispute these numbers, saying they are inflated in order to attract more funding from the Costa Rican government and outside sources.

8. Tomatoes are not a cucurbit.

9. This is partly because many minivegetable farmers purchase seeds from the minivegetable exporters, while chayote farmers have control over the germplasm, selecting their seed from their chayote fruit with desirable characteristics.

10. SuperChayote pays for residue tests of its produce from its farming operation about every three months. The manager reported that the company had never had any residue problems.

11. Methamidophos is the most commonly found illegal residue on produce imported into the United States. Between 1996 and 2006, methamidophos caused 16 percent of all residue violations on imported fresh vegetables while only accounting for 0.4 percent of residue tests (Galt 2010).

12. By "rationalization," used here and in following chapters, I do not mean pesticide-use reduction per se but rather the process by which farmers use pesticides according to their labels and attempt to lessen impacts for consumers. In many cases of individual farmers, rationalization of pesticide use does indeed involve reducing pesticide use overall.

Chapter 5

Reprinted with modifications from Ryan E. Galt, "Regulatory Risk and Farmers' Caution with Pesticides in Costa Rica," *Transactions of the Institute of British Geographers* 32, no. 3 (2007): 377–94. Reprinted with permission of the author.

1. State regulation remains more important in the United States than in the United Kingdom, where much agrifood regulation is now conducted by supermarkets (Flynn, Marsden, and Whatmore 1994).

2. Many developing countries have these types of regulations, but the lack of enforcement and punitive measures means that farmers face little risk in not complying, as we will see in chapter 6.

3. Some data are from the very large farming operations of the exporting firms, but most are from small- to medium-scale farmers who sell directly to the exporters.

4. The reason for lack of registration is difficult to determine. It can mean that registration failed presumably because of health or environmental risks or that the firm that owns the patent has not attempted registration.

5. Another contradiction arises with this use of low doses: while it works for residues, low doses can also speed the development of pest resistance.

6. The survey did not include questions about label interpretation, so the extent of this interpretation cannot be determined. Similar views, however, did come up in other surveys and interviews.

7. Specifically, a correlation between the two variables—sum of dangers known and sum of protective gear worn—yields a correlation value of $r = -0.01$, where $r = 1$ or $r = -1$ is a perfect association and $r = 0$ is a complete lack of relationship (Galt 2013).

Chapter 6

Reprinted with modifications from Ryan E. Galt, "'It Just Goes to Kill Ticos': National Market Regulation and the Political Ecology of Farmers' Pesticide Use in Costa Rica," *Journal of Political Ecology* 16 (2009): 1–33. Reprinted with permission of the author.

1. Shortly after the introduction of pesticides in India, one hundred people died in Kerala after eating parathion-contaminated wheat flour (Gupta 2004, 85). Many other such deaths have been reported in other countries (Dinham 1993, 55). Historically, there were cases of children in the United States being killed by lead arsenate residues on produce (Whorton 1974). It is conceivable that lead arsenate still causes such problems where it is used but that these go undocumented.

2. Poisoning outbreaks from pesticide residues have also happened in the United States recently (Farley and McFarland 1999; Green et al. 1987) and historically (Whorton 1974).

3. Food safety issues in Costa Rica extend far beyond a lack of control over farmers' pesticide use. Of the food industry businesses surveyed, 63 percent have "no structured sanitation or cleaning program," 78 percent "have no hand cleaning facilities for employees," and 67 percent "do not control or treat water used for processing" (Kopper 2002, 3).

4. The implication is that the samples are not randomly and independently chosen, which could mean that sellers are able to provide produce from their most trusted farmers.

5. Valverde G., Carazo Rojas, and Araya Rojas (2001) found that both marketplaces had residues above tolerances at a similar rate for sweet peppers, leading the authors to conclude that marketplace makes no difference for pesticide residues. For fecal coliform contamination, however, they found that supermarkets tend to have lower levels of contamination, though "both sites appear to have fecal contamination problems with their products" (65). Fecal contamination became a national concern after a television news program showed that lettuce grown near Cartago was irrigated with water from drainage systems that receive untreated sewage.

6. Presumably Opondo (2000) means the pyrethroids, which are perhaps less environmentally damaging than other types of insecticides since they break down faster. It is important to note, however, that they still can impact fish and wildlife before they decompose and have negative human health effects.

Conclusion

1. An example of this strong motivation without a solid economic foundation is that some farmers who have experienced intense episodes of pesticide poisoning have attempted to reduce their pesticide use, but certainly not all of them have done so.

Appendix 1

1. Three of the 148 surveys did not yield adequate pesticide-use information for field-specific crop-spraying schedules (defined in the main text). Other information from these three surveys is useful and is used for other types of data analysis.

Glossary of Pesticide Terms

active ingredient (ai): The chemical compound that acts against the pest organism. As a measurement, it is used to standardize comparisons of pesticide intensity. *See also* pesticide intensity.

bad actor pesticides: A classification created by the Pesticide Action Network (PAN), an activist and research organization, to differentiate between pesticides known to cause problems and those that are likely to be less harmful. A pesticide qualifies as a bad actor if it is any of the following: a highly acute toxin according to the World Health Organization, the U.S. Environmental Protection Agency (EPA), or the U.S. National Toxicology Program; a known or probable carcinogen according to the EPA; a reproductive or developmental toxin listed in California's Proposition 65 list; a cholinesterase inhibitor according to the Material Safety Data Sheet, the California Department of Pesticide Regulation, or the PAN staff's evaluation of chemical structure; or a known groundwater contaminant.

biopesticides: Pesticides of direct biological origin, such as plant extracts.

carbamates: A group of synthetic pesticides that reversibly inactivate the acetylcholinesterase enzyme, which is a key component of the nervous system (the organophosphates irreversibly block the same enzyme). Like the organophosphates, these compounds are acutely toxic to humans and other animals, meaning that contact with small doses can lead to poisoning or death.

formulated product: Consists of the active ingredient and other ingredients, such as surfactants and emulsifiers, that help the active ingredient or are used to improve storage, application, and/or handling.

inorganic pesticides: Pesticide compounds without carbon, typically created from metallic elements such as arsenic, copper, and lead as well as nonmetallic elements such as sulfur. Their use predates synthetic pesticides made from the tools of organic chemistry.

integrated pest management (IPM): Originally developed as an alternative to heavy pesticide spraying with an emphasis on control of pest organisms through cultural practices (timing, rotations, cultivation, variety selection, etc.) and biological control (promoting conditions that supported beneficial insects such as predators and parasites of pests). The concept relies on economic thresholds, that is, not spraying until it makes definitive economic sense to do so, which demands data collection on pest populations. The term has been co-opted by industry and is sometimes critiqued as "integrated pesticide management."

LD_{50}: Stands for "lethal dose, 50 percent" and means the dose at which a compound kills half of the test organisms during a controlled toxicological test. LD_{50} is a measurement commonly used in toxicology to assess the acute toxicity of compounds.

maximum residue limit (MRL): *See* **tolerance.**

organochlorines: A group of synthetic pesticides that were very efficacious in agriculture because of their persistence. These were the first popular synthetic pesticides, adopted widely after World War II. DDT is the most (in)famous of this group. Generally, these compounds persist in the environment and bioaccumulate, increasing in concentration in fat tissue in organisms higher on the food chain. Their threats to humans and wild animals led to most of them being banned in the industrialized world in the 1970s.

organophosphates: A group of synthetic pesticides and nerve agents that irreversibly block the acetylcholinesterase enzyme, required for nerve functioning in insects, humans, and other animals. The organophosphates typically degrade faster than the organochlorine pesticides but are much more acutely toxic. These pesticides usually cause the most poisoning events among farmers and farmworkers in the developing world. Organophosphates are particularly hazardous to the brain development of fetuses and children. Organophosphate compounds include the nerve agent sarin, classified as a weapon of mass destruction.

parts per million (ppm): Equivalent to milligrams per kilogram and commonly used as a measurement to specify residues detected (on food or in the environment), tolerances, and LD_{50}.

pesticide intensity: The amount of pesticide used over a spatial area and a time horizon (e.g., active ingredient per hectare per year, which I write as ai/ha/yr).

preharvest interval (PHI): Refers to the amount of time between a pesticide application and the harvest. Generally, the shorter the PHI on a pesticide label, the more rapid the degradation of a pesticide. Pesticides with very low PHIs (such as pyrethroids) often degrade rapidly, even below the detection limit of sophisticated residue testing equipment.

pyrethroids: A group of synthetic insecticides modeled after pyrethrin, a biopesticide from chrysanthemums. Pyrethroids are generally applied at much lower doses per hectare than the organophosphate, carbamate, and organochloride pesticides and are generally less acutely toxic than organophosphates and carbamates. Unlike these two groups, pyrethroids are not acetylcholinesterase inhibitors. They break down very rapidly in sunlight and the atmosphere. Despite these beneficial properties, they are toxic to beneficial insects and acutely toxic to aquatic organisms.

synthetic pesticides: Pesticide compounds made from organic chemicals (sharing the meaning of organic chemistry) that contain carbon and are derived from hydrocarbons found in fossil fuels. In common parlance, these compounds of organic chemistry are synthetic. Their mass production and adoption in agriculture and pest control occurred in the 1940s and 1950s, starting with the organochlorines.

tolerance: The level of a substance, including pesticide residues, permitted on foods according to various food standards. The term "tolerance" is somewhat confusing, since it implies that the set level is the level at which our bodies can tolerate a compound, yet Wargo (1998, 315n25) notes that "in reality few tolerances were established to be health protective. It really means the level of contamination that is legally tolerated."

References

Abad, Z. G., and J. A. Abad. 1995. "Historical Evidence on the Occurrence of Late Blight of Potato, Tomato and Pearl Melon in the Andes of South America." In *Phytophthora Infestans 150*, edited by L. J. Dowley, E. Bannon, L. R. Cooke, T. Keane, and E. O'Sullivan, 36–41. Dublin, Ireland: Boole.

Abelson, Philip H. 1994. "Risk Assessment of Low-level Exposures." *Science* 265, no. 5176: 1507.

Abeysekera, W. A. T. 1988. "Pesticide Use in the Food Production Sector in Sri Lanka." In *Use of Pesticides and Health Hazards in the Plantation Sector*, edited by Congress Labour Foundation, 12–37. Colombo, Sri Lanka: Friedrich-Ebert-Stiftung.

Agne, Stefan, Hermann Waibel, and Octavio Ramírez. 1999. "Diagnóstico y recomendaciones sobre criterios económicos y legislación para el uso de plaguicidas en Costa Rica." *Manejo Integrado de Plagas*, no. 54: 44–52.

Agrawal, Arun. 2005. *Environmentality: Technologies of Government and the Making of Subjects*. Durham, NC: Duke University Press.

Agriculture and Agri-Food Canada. 2012. "Agri-Food Trade Policy Negotiations to Modernize the Canada—Costa Rica Free Trade Agreement." Agriculture and Agri-Food Canada, http://www.agr.gc.ca/itpd-dpci/ag-ac/4958-eng.htm.

Aguilar-del Real, Ana, Antonio Valverde-García, and Francisco Camacho-Ferre. 1999. "Behavior of Methamidophos Residues in Peppers, Cucumbers, and Cherry Tomatoes Grown in a Greenhouse: Evaluation by Decline Curves." *Journal of Agricultural and Food Chemistry* 47: 3355–58.

Aistara, Guntra A. 2011. "Seeds of Kin, Kin of Seeds: The Commodification of Organic Seeds and Social Relations in Costa Rica and Latvia." *Ethnography* 12, no. 4: 490–517.

Albert, Lilia A. 2005. "Panorama de los plaguicidas en México." Paper presented at the 7th Congreso de Actualización en Toxicología Clínica, Nayarit, Mexico.

Altieri, Miguel A. 2000. "Ecological Impacts of Industrial Agriculture and the Possibilities for Truly Sustainable Farming." In *Hungry for Profit: The Agribusiness Threat to Farmers, Food, and the Environment*, edited by F. Magdoff, J. B. Foster, and F. H. Buttel, 77–92. New York: Monthly Review Press.

Altieri, Miguel A., and Clara I. Nicholls. 2003. "Soil Fertility Management and Insect Pests: Harmonizing Soil and Plant Health in Agroecosystems." *Soil & Tillage Research* 72: 203–11.

Amekawa, Yuichiro. 2013. "Can a Public GAP Approach Ensure Safety and Fairness? A Comparative Study of Q-GAP in Thailand." *Journal of Peasant Studies* 40, no. 1: 189–217.

Ames, Bruce N., Margie Profet, and Lois Swirsky Gold. 1990. "Dietary Pesticides (99.99% All Natural)." *Proceedings of the National Academy of Sciences of the United States of America* 87, no. 19: 7777–81.

Andreatta, Susan L. 1997. "Bananas, Are They the Quintessential Health Food? A Global/Local Perspective." *Human Organization* 56, no. 4: 437–49.

———. 1998a. "Agrochemical Exposure and Farmworker Health in the Caribbean: A Local/Global Perspective." *Human Organization* 57, no. 3: 350–58.

———. 1998b. "Transformation of the Agro-Food Sector: Lessons from the Caribbean." *Human Organization* 57, no. 4: 414–29.

Andréu, Tomás. 2011. "Costa Rica: Country Tops Farm Chemical Use List." *Latin America Press,* September 1.

Anonymous. 1988a. "Agroquímicos y exportación." *La República,* March 6, 1988, 10.

———. 1988b. "No se ha paralizado la exportación de chayote." *La República,* March 5, 1988, 9.

———. 1988c. "Suspenden compra de chayotes a C.R." *La República,* March 4, 1988, 2.

———. 1998. "Análisis de residuos de plaguicidas en seis diferentes hortalizas de consumo nacional procedentes de diversas regiones de Nicaragua." *For Export (Nicaragua)* (September–October): 29–31.

———. 2008a. "Agricultores protestaron contra altos precios de plaguicidas." *Al Día,* April 22.

———. 2008b. "Melón y sandía superan controles internacionales." *Al Día,* July 16.

Antle, John M., Donald C. Cole, and Charles C. Crissman. 1998. "Further Evidence on Pesticides, Productivity and Farmer Health: Potato Production in Ecuador." *Agricultural Economics* 18, no. 2: 199–207.

Apple, A. E., and W. E. Fry. 1983. *Potato Late Blight* (Phytophthora infestans) *(Mont.).* Ithaca, NY: Cooperative Extension, Cornell University, Department of Plant Pathology.

Arauz C., Luis Felipe, Elizabeth Carazo R., and Dennis Mora A. 1983. "Diagnóstico sobre el uso y manejo de plaguicidas en las fincas hortícolas del Valle Central de Costa Rica." *Agronomía y Ciencia* 1, no. 3: 37–49.

Arbeláez, Patricia, and Samuel Henao H. 2002. *Situación epidemiológica de las intoxicaciones agudas por plaguicidas en el istmo centroamericano.* San José, Costa Rica: PLAGSALUD.

Arbona, Sonia I. 1998. "Commercial Agriculture and Agrochemicals in Almolonga, Guatemala." *Geographical Review* 88, no. 1: 47–63.

Arias Rodríguez, Fabio. 1998. "Evaluación económica de las diversas dosis y frecuencias de aplicación de plaguicidas en papa y cebolla utilizados por un grupo de agricultores en la Zona de Tierra Blanca de Cartago." Unpublished thesis, Economía Agrícola, Universidad de Costa Rica, San José, Costa Rica.

Arnáez, Elizabeth, Hilda Quesada, Enrique Hernández, Virginia Valverde, and Benjamín Mora. 1993. "Uso y manejo de plaguicidas en El Valle de El Guarco, Cartago." *Tecnología en Marcha* 12, no. 2: 51–59.

Arrieta Chavarría, Omar. 1984. "La organización del espacio en sociedades agrarias: El caso de Cot-Irazú." Unpublished thesis, Facultad de Ciencias Sociales, Escuela de Historia y Geografía, Universidad de Costa Rica, San José, Costa Rica.

Avery, Dennis T. 1995. *Saving the Planet with Pesticides and Plastic: The Environmental Triumph of High-Yield Farming.* Indianapolis: Hudson Institute.

Baker, B. P., C. M. Benbrook, Edward Groth III, and K. Lutz Benbrook. 2002. "Pesticide Residues in Conventional, Integrated Pest Management (IPM)-Grown and Organic Foods: Insights from Three US Data Sets." *Food Additives and Contaminants* 19, no. 5: 427–46.

Balsevich, F., J. Berdegué, L. Flores, D. Mainville, T. Reardon, L. Busch, and L. Unnevehr. 2003. "Supermarkets and Produce Quality and Safety Standards in Latin America." *American Journal of Agricultural Economics* 85, no. 5: 1147–54.

Banaji, Jarius. 1980. "Summary of Selected Parts of Kautsky's *The Agrarian Question.*" In *The Rural Sociology of Advanced Societies: Critical Perspectives,* edited by F. H. Buttel and H. Newby, 39–82. Montclair, NJ: Allanheld, Osmun.

Barbosa, S. 2000. "IPM for Vegetables in Latin America: Problems and Perspectives." *Acta Horticulturae* 513: 105–7.

Barham, Bradford L., Michael R. Carter, and Wayne Sigelko. 1995. "Agro-Export Production and Peasant Land Access: Examining the Dynamic between Adoption and Accumulation." *Journal of Development Economics* 46, no. 1: 85–107.

Barham, Bradford L., Mary Clark, Elizabeth Katz, and Rachel Schurman. 1992. "Nontraditional Agricultural Exports in Latin America." *Latin American Research Review* 27, no. 2: 43–82.

Barnes, Trevor. 1984. "Theories of Agricultural Rent within the Surplus Approach." *International Regional Science Review* 9, no. 2: 125–40.

Barnett, John B., Theo Colborn, Michel Fournier, John Gierthy, Keith Grasman, Nancy Kerkvliet, Garet Lahvis, Michael Luster, Peter McConnachie, J. Peterson Myers, A. D. M. E. Osterhaus, Robert Repetto, Rosalind Rolland, Louise Rollins-Smith, Ralph Smialowicz, Michael Smolen, Sara Walker, and David Watkins. 1996. "Statement from the Work Session on Chemically-Induced Alterations in the Developing Immune System: The Wildlife/Human Connection." *Environmental Health Perspectives* 104 (Supplement 4): 807–8.

Barquero S., Marvin. 1990. "MAG niega peligro por agroquímicos." *La Nación,* November 15, 6A.

————. 1999. "Guerra a los plaguicidas." *La Nación,* April 24, 25A.

Barquero S., Marvin, and Patricia Navarro. 1990a. "Anarquía en control de agro-químicos: Los ministerios de Salud y de Agricultura no se ponen de acuerdo sobre el registro de plaguicidas." *La Nación,* November 15, 5A.

————. 1990b. "Residuos superan límites en muestras de vegetales: Ocho de nueve muestras de vegetales analizadas, con niveles altos de residuos de agroquímicos, procedían de Cartago." *La Nación,* November 13, 6A.

Barrantes F., Jorge Arturo, Alfonso Liao, and Amán Rosales. 1985. *Atlas climatológico de Costa Rica.* San José, Costa Rica: Ministerio de Agricultura y Ganadería and Instituto Meteorológico Nacional.

Barrett, Hazel R., Brian W. Ilbery, Angela W. Browne, and Tony Binns. 1999. "Globalization and the Changing Networks of Food Supply." *Transactions of the Institute of British Geographers* 24: 159–74.

Barry, Tom. 1987. *Roots of Rebellion: Land and Hunger in Central America.* Boston: South End.

Bartlett, Andrew, and Hein Bijlmakers. 2003. *Did You Take Your Poison Today?* Bangkok: IPM DANIDA.

Bassett, Thomas J. 1988. "The Political Ecology of Peasant-Herder Conflicts in the Northern Ivory Coast." *Annals of the Association of American Geographers* 78, no. 3: 453–72.

Beck, Ulrich. 1992. *The Risk Society: Towards a New Modernity.* Translated by M. Ritter. Newbury Park, CA: Sage.

Bee, Anna. 2000. "Globalization, Grapes, and Gender: Women's Work in Traditional and Agro-export Production in Northern Chile." *Geographical Journal* 166, no. 3: 255–65.

Bell, Erin M., Irva Hertz-Picciotto, and James J. Beaumont. 2001. "Case-Cohort Analysis of Agricultural Pesticide Applications near Maternal Residence and Selected Causes of Fetal Death." *American Journal of Epidemiology* 154, no. 8: 702–10.

Bell, Michael. 2004. *Farming for Us All: Practical Agriculture and the Cultivation of Sustainability.* University Park: Pennsylvania State University Press.

Benbrook, Charles M. 1997. *Steering Clear of the Pesticide Treadmill.* Paper given at the Northern Plains Sustainable Agriculture Society, January 18, http://www.pmac .net/steering.htm.

————. 2002. "Organochlorine Residues Pose Surprisingly High Dietary Risks." *Journal of Epidemiology and Community Health* 56: 822–23.

Berdegué, Julio A., Fernando Balsevich, Luis Flores, and Thomas Reardon. 2005. "Central American Supermarkets' Private Standards of Quality and Safety in Procurement of Fresh Fruits and Vegetables." *Food Policy* 30, no. 3: 254–69.

Bernhardt, Ed, J. Dodson, and J. C. Watterson. 1988. *Cucurbit Diseases: A Practical Guide for Seedsmen, Growers and Agricultural Advisors.* Saticoy, CA: Petoseed Co.

Bingen, R. James, and Lawrence Busch, eds. 2006. *Agricultural Standards: The Shape of the Global Food and Fiber System.* Dordrecht: Springer.

Birnbaum, Linda S. 1994. "Endocrine Effects of Prenatal Exposure to PCBs, Dioxins, and Other Xenobiotics: Implications for Policy and Future Research." *Environmental Health Perspectives* 102, no. 8: 676–79.

Bischoff, Robert F. 1993. "Pesticide Chemicals: An Industry Perspective on Minor Crop Uses." In *New Crops*, edited by J. Janick and J. E. Simon, 662–64. New York: Wiley.

Blaikie, Piers. 1985. *The Political Economy of Soil Erosion in Developing Countries.* London: Methuen.

———. 1988. "The Explanation of Land Degradation in Nepal." In *Deforestation: Social Dynamics in Watersheds and Mountain Ecosystems,* edited by J. Ives and D. C. Pitt, 132–58. London: Routledge.

Blaikie, Piers, and Harold Brookfield. 1987. *Land Degradation and Society.* London: Methuen.

Blanco, H., P. J. Shannon, and J. L. Saunders. 1990. "Resistencia de *Plutella xylostella* (Lep: *Plutellidae*) a tres piretroides sintéticos en Costa Rica." *Turrialba* 40, no. 2: 159–64.

Block, Fred. 2003. "Karl Polanyi and the Writing of *The Great Transformation.*" *Theory and Society* 32, no. 3: 275–306.

Boh, Cheah Uan. 2003. "MRLs and International Trade: A Developing Country Perspective." In *Chemistry of Crop Protection: Progress and Prospects in Science and Regulation,* edited by G. Voss and G. Ramos, 371–81. Cambridge: Wiley-VCH.

Bojacá, Carlos Ricardo, Luis Alejandro Arias, Diego Alejandro Ahumada, Héctor Albeiro Casilimas, and Eddie Schrevens. 2013. "Evaluation of Pesticide Residues in Open Field and Greenhouse Tomatoes from Colombia." *Food Control* 30, no. 2: 400–403.

Bolaños, Rafael, Miriam Quirós, Sergio Alvarado, Emma Rosa Quirós, Claudio Orozco, Joaquín Solano, and Claudio Rojas. 1993. *Ayer Ujarrás . . . hoy Paraíso: Siglos XVI–XX.* San José, Costa Rica: Lithocolor.

Bolaños Herrera, Alfredo. 2001. *Introducción a la Olericultura.* San José, Costa Rica: EUNED.

Born, Branden, and Mark Purcell. 2006. "Avoiding the Local Trap: Scale and Food Systems in Planning Research." *Journal of Planning Education and Research* 26, no. 2: 195–207.

Bosa, Carlos Felipe, Alba Marina Cotes Prado, Takehiko Fukumoto, Marie Bengtsson, and Peter Witzgall. 2005. "Pheromone-mediated communication disruption in Guatemalan potato moth, *Tecia solanivora.*" *Entomologia Experimentalis et Applicata* 114, no. 2: 137–42.

Boyd, William, W. S. Prudham, and R. A. Schurman. 2001. "Industrial Dynamics and the Problem of Nature." *Society & Natural Resources* 14, no. 7: 555–70.

Brenes, Lidiette. 1991. "Pérdidas millonarias por rechazo de exportaciones." *La Nación,* August 19, 5A.

Breslin, Patrick. 1996. "Costa Rican Farmers Find Their Mini-niche." *Grassroots Development* 20, no. 2: 26–33.

Brethour, Cher, and Alfons Weersink. 2001. "An Economic Evaluation of the Environmental Benefits from Pesticide Reduction." *Agricultural Economics* 25: 219–26.

Brohman, John. 1996. "The Agroexport Model and Nontraditional Exports in Central America: Déjà Vu or Something New?" *Yearbook, Conference of Latin Americanist Geographers* 22: 1–16.

Brown, J. Christopher, and Mark Purcell. 2005. "There's Nothing Inherent about Scale: Political Ecology, the Local Trap, and the Politics of Development in the Brazilian Amazon." *Geoforum* 36, no. 5: 607–24.

Bryant, Raymond L., and Sinéad Bailey. 1997. *Third World Political Ecology.* New York: Routledge.

Buhs, Joshua Blu. 2002. "Dead Cows on a Georgia Field: Mapping the Cultural Landscape of the Post–World War II American Pesticide Controversies." *Environmental History* 7, no. 1: 99–121.

Bull, David. 1982. *A Growing Problem: Pesticides and the Third World Poor.* Oxford: OXFAM.

Bullard, Robert D. 1993. "Anatomy of Environmental Racism and the Environmental Justice Movement." In *Confronting Environmental Racism: Voices from the Grassroots,* edited by R. D. Bullard, 15–39. Boston: South End.

Bunch, Roland. 1982. *Two Ears of Corn: A Guide to People-Centered Agricultural Improvement.* 2nd ed. Oklahoma City: World Neighbors.

Burawoy, Michael. 2000. "Grounding Globalization." In *Global Ethnography: Forces, Connections, and Imaginations in a Postmodern World,* edited by M. Burawoy, 337–50. Berkeley: University of California Press.

Busch, Lawrence. 2011. *Standards: Recipes for Reality.* Cambridge, MA: MIT Press.

Buttel, Frederick H. 1983. "Farm Structure and Rural Development." In *Farms in Transition: Interdisciplinary Perspectives on Farm Structure,* edited by D. E. Brewster, W. D. Rasmussen, and G. Youngberg, 103–24. Ames: Iowa State University Press.

———. 2006. "Sustaining the Unsustainable: Agro-food Systems and Environment in the Modern World." In *Handbook of Rural Studies,* edited by P. J. Cloke, T. Marsden and P. H. Mooney, 213–29. Thousand Oaks, CA: Sage.

Calderón Coto, Oscar R. 1972. *Manual de costos básicos de actividades agropecuarias, 1972.* Cartago, Costa Rica: Banco Crédito Agrícola de Cartago.

Calderón Mata, Saúl. 1980. "Tecnología empleada por productores del cantón de Paraíso en el cultivo de chayote y apio." Unpublished thesis, Departamento de Agronomía, Instituto Tecnológico de Costa Rica, Cartago, Costa Rica.

Capinera, John L. 2001. *Handbook of Vegetable Pests.* San Diego: Academic Press.

Carazo, Elizabeth, Manuel Constenla, Gilbert Fuentes, and Pran Nath Moza. 1984. "Studies of Methamidophos-C-14 in Costa Rican Vegetables and Soils." *Chemosphere* 13, no. 8: 939–46.

Carletto, Calogero, Alain de Janvry, and Elisabeth Sadoulet. 1999. "Sustainability in the Diffusion of Innovations: Smallholder Nontraditional Agro-Exports in Guatemala." *Economic Development and Cultural Change* 47, no. 2: 345–69.

Carson, Rachel. 1994. *Silent Spring.* New York: Houghton Mifflin.

Carter, Michael R., and Bradford L. Barham. 1996. "Level Playing Fields and *Laissez faire:* Postliberal Development Strategy in Inegalitarian Agrarian Economies." *World Development* 24, no. 7: 1133–49.

Carter, Michael R., Bradford L. Barham, and Dina Mesbah. 1995. "Agricultural Export Booms and the Rural Poor in Chile, Guatemala, and Paraguay." *Latin American Research Review* 31, no. 1: 33–65.

Cartín L., Victor M., Elizabeth Carazo R., Jorge A. Lobo S., Luis A. Monge V., and Lizbeth Araya R. 1999. "Resistencia de *Plutella xylostella* a *Bacillus thuringiensis* en Costa Rica." *Manejo Integrado de Plagas,* no. 54: 31–36.

Castellanos Robayo, Armando. 1972. "Evaluación de los suelos de la zona este del cantón de Oreamuno." Unpublished hesis, Ingeniero Agrónomo, Facultad de Agronomía, Universidad de Costa Rica, San José, Costa Rica.

Castillo, Luisa E., Elba de la Cruz, and Clemens Ruepert. 1997. "Ecotoxicology and Pesticides in Tropical Aquatic Ecosystems of Central America." *Environmental Toxicology and Chemistry* 16, no. 1: 41–51.

Castillo Nieto, Silvia. 1999. "Sello azul certifica control de plaguicidas." *El Financiero,* May 31, 17.

Castleman, B. I., and V. Navarro. 1987. "International Mobility of Hazardous Products, Industries, and Wastes." *Annual Review of Public Health* 8: 1–19.

Castree, Noel, and Bruce Braun. 1998. "The Construction of Nature and the Nature of Construction: Analytical and Political Tools for Building Survivable Futures." In *Remaking Reality: Nature at the Millennium,* edited by B. Braun and N. Castree, 3–42. New York: Routledge.

Ceciliano Romero, Antonio. 1986. "Análisis en el manejo racional de agroquímicos, prácticas culturales y agronómicas, asistencía técnica directa y extensión agronómica en el cultivo del café en el sur de Cartago y San José." Unpublished thesis, Departamento de Administración de Empresas Agropecuarias, Instituto Tecnológico de Costa Rica, Cartago, Costa Rica.

CentralAmericaData. 2010. "Costa Rica: Industry Want Duty-free Potatoes." CentralAmericaData, November 8, http://www.centralamericadata.com/en/article/home/Costa_Rica_Industry_Want_Dutyfree_Potatoes.

Centre for Science and Environment. 2006. *Poison vs Nutrition: A Briefing Paper on Pesticide Contamination and Food Safety.* New Delhi: Centre for Science and Environment.

Centro Internacional de la Papa. n.d. "The World Potato Atlas: Costa Rica." Centro Internacional de la Papa, https://research.cip.cgiar.org/confluence/display/wpa/Costa+Rica.

Chan, Thomas Y. K. 2001. "Vegetable-borne Methamidophos Poisoning (Letter)." *Clinical Toxicology* 39, no. 4: 337–38.

Chang, Ju-Mei, Tay-Hwa Chen, and Tony J. Fang. 2005. "Pesticide Residue Monitoring in Marketed Fresh Vegetables and Fruits in Central Taiwan (1999–2004) and an Introduction to the HACCP System." *Journal of Food and Drug Analysis* 13, no. 4: 368–76.

Chaverri, Fabio. 1999. *Importación y uso de plaguicidas en Costa Rica: Análisis del período, 1994–1996*. Heredia, Costa Rica: Instituto Regional de Estudios en Sustancias Tóxicas, Universidad Nacional de Costa Rica.

Chávez, César. 1993. "Farm Workers at Risk." In *Toxic Struggles: The Theory and Practice of Environmental Justice*, edited by R. Hofrichter, 163–70. Philadelphia: New Society Publishers.

Chayanov, Alexander V. 1966. *The Theory of Peasant Economy*. Edited by D. Rhorner, B. H. Kerblay and R. E. F. Smith. Homewood, IL: Published for the American Economic Association by R. D. Irwin.

Clapp, Roger Alex. 1994. "The Moral Economy of the Contract." In *Living under Contract: Contract Farming and Agrarian Transformation in Sub-Saharan Africa*, edited by P. D. Little and M. J. Watts, 78–94. Madison: University of Wisconsin Press.

Clark, Mary A. 1995. "Nontraditional Export Promotion in Costa Rica: Sustaining Export-Led Growth." *Journal of Interamerican Studies and World Affairs* 37, no. 2: 181–223.

Cochrane, Willard W. 1979. *The Development of American Agriculture: A Historical Analysis*. Minneapolis: University of Minnesota Press.

Cochrane, Willard Wesley, and C. Ford Runge. 1992. *Reforming Farm Policy: Toward a National Agenda*. 1st ed. Ames: Iowa State University Press.

Colborn, Theo. 1995. "Environmental Estrogens: Health Implications for Humans and Wildlife." *Environmental Health Perspectives* 103, Supplement 7: 135–36.

Colborn, Theo, Dianne Dumanoski, and John Peterson Myers. 1997. *Our Stolen Future: Are We Threatening Our Fertility, Intelligence, and Survival? A Scientific Detective Story*. New York: Penguin.

Colborn, Theo, Frederick S. vom Saal, and Ana M. Soto. 1993. "Developmental Effects of Endocrine-Disrupting Chemicals in Wildlife and Humans." *Environmental Health Perspectives* 101, no. 5: 378–84.

Collins, Jane L. 1986. "Smallholder Settlement of Tropical South America: The Social Causes of Ecological Destruction." *Human Organization* 45, no. 1: 1–10.

———. 1995. "Farm Size and Non Traditional Exports: Determinants of Participation in World Markets." *World Development* 23, no. 7: 1103–14.

Collins, Jane L., and Greta R. Krippner. 1999. "Permanent Labor Contracts in Agriculture: Flexibility and Subordination in a New Export Crop." *Comparative Studies in Society and History* 41, no. 3: 510–34.

Comité Técnico de *Liriomyza*. 1990. *El "Minador de las hojas" Liriomyza sp. (Diptera; Agromyzidae)*. Ministerio de Agricultura y Ganadería, CATIE, Convenio Costarricense Alemán, Sanidad Vegetal-GTZ.

Conroy, Michael E., Douglas L. Murray, and Peter R. Rosset. 1996. *A Cautionary Tail: Failed U.S. Development Policy and Central America*. Boulder, CO: Lynne Rienner.

CropLife International. 2007. *Annual Report 2006/2007*. Brussels: CropLife International.

———. 2010. *Facts and Figures: The Status of Global Agriculture*. CropLife International, www.croplife.org/view_document.aspx?docId=2877.

Crosby, Alfred W. 1990. "An Enthusiastic Second." *Journal of American History* 76, no. 4: 1107–10.

Cupples, Julie, and Irving Larios. 2010. "A Functional Anarchy: Love, Patriotism, and Resistance to Free Trade in Costa Rica." *Latin American Perspectives* 37, no. 6: 93–108.

Damiani, Octavio. 2003. "Effects on Employment, Wages, and Labor Standards of Non-traditional Export Crops in Northeast Brazil." *Latin American Research Review* 38, no. 1: 83–112.

Darby, William J. 1962. "Silence, Miss Carson." *Chemical and Engineering News* 40: 60–63.

Dark, Petra, and Henry Gent. 2001. "Pests and Diseases of Prehistoric Crops: A Yield 'Honeymoon' for Early Grain Crops in Europe?" *Oxford Journal of Archaeology* 20, no. 1: 59–78.

Davis, John Emmeus. 1980. "Capitalist Agricultural Development and the Exploitation of the Propertied Laborer." In *The Rural Sociology of the Advanced Societies: Critical Perspectives,* edited by F. H. Buttel and H. Newby, 133–53. London: Croom Held.

Davis, Mike. 2001. *Late Victorian Holocausts: El Niño Famines and the Making of the Third World*. New York: Verso.

de Janvry, Alain. 1981. *The Agrarian Question and Reformism in Latin America*. Baltimore: Johns Hopkins University Press.

———. 1985. "Social Disarticulation in Latin American History." In *Debt and Development in Latin America,* edited by K. S. Kim and D. F. Ruccio, 32–73. Notre Dame, IN: University of Notre Dame Press.

DeLind, Laura B. 1991. "Sustainable Agriculture in Michigan: Some Missing Dimensions." *Agriculture and Human Values* 8, no. 4: 38–45.

Díaz-Knauf, Katherine, Carmen Ivankovich, Fernando Aguilar, Christine Bruhn, and Howard Schutz. 1993. "Consumer Attitudes toward Food Safety of Produce in Costa Rica." *Journal of Foodservice Systems* 7, no. 2: 105–15.

Dich, Jan, Shelia Hoar Zahm, Annika Hanberg, and Hans-Olov Adami. 1997. "Pesticides and Cancer." *Cancer Causes and Control* 8, no. 3: 420–43.

Dinham, Barbara. 1991. "FAO and Pesticides: Promotion or Proscription?" *Ecologist* 21, no. 2 (March–April): 61–65.

———. 1993. *The Pesticide Hazard: A Global Health and Environmental Audit*. Atlantic Highlands, NJ: Zed Books.

———. 2003. "Growing Vegetables in Developing Countries for Local Urban Populations and Export Markets: Problems Confronting Small-Scale Producers." *Pest Management Science* 59, no. 5: 575–82.

Dirección General de Estadística y Censos. 1953. *Censo agropecuario de 1950.* San José, Costa Rica: Impreso en el Instituto Geográfico.

———. 1965. *Censo agropecuario, 1963.* San José, Costa Rica: Dirección General de Estadística y Censos, Sección de Publicaciones.

———. 1986. *Censo agropecuario, 1984.* San José, Costa Rica: Ministerio de Economía, Industria y Comercio, Dirección General de Estadística y Censos.

Dogheim, Salwa M., Eslam N. Nasr, Monir M. Almaz, and Mahmoud M. El-Tohamy. 1990. "Pesticide Residues in Milk and Fish Samples Collected from Two Egyptian Governates." *Journal of the Association of Official Analytical Chemists* 73, no. 1: 19–21.

Dolan, C., and J. Humphrey. 2000. "Governance and Trade in Fresh Vegetables: The Impact of UK Supermarkets on the African Horticulture Industry." *Journal of Development Studies* 37, no. 2: 147–76.

Dugger, Celia W. 2004. "Supermarket Giants Crush Central American Farmers." *New York Times,* December 28.

Durham, William H. 1995. "Political Ecology and Environmental Destruction in Latin America." In *The Social Causes of Environmental Destruction in Latin America,* edited by M. Painter and W. H. Durham, 249–64. Ann Arbor: University of Michigan Press.

Ecobichon, Donald J. 2001. "Pesticide Use in Developing Countries." *Toxicology* 160, nos. 1–3: 27–33.

Edelman, Marc. 1989. "Illegal Renting of Agrarian Reform Plots: A Costa Rican Case Study." *Human Organization* 48, no. 2: 172–79.

———. 1995. "Rethinking the Hamburger Thesis: Deforestation and the Crisis of Central America's Beef Exports." In *The Social Causes of Environmental Destruction in Latin America,* edited by M. Painter and W. H. Durham, 25–62. Ann Arbor: University of Michigan Press.

———. 1999. *Peasants against Globalization: Rural Social Movements in Costa Rica.* Stanford, CA: Stanford University Press.

Edwards, Clive A. 1993. "The Impact of Pesticides on the Environment." In *The Pesticide Question: Environment, Economics, and Ethics,* edited by D. Pimentel and H. Lehman, 13–46. New York: Chapman and Hall.

El Sebae, A. H. 1993. "Special Problems Experienced with Pesticide Use in Developing Countries." *Regulatory Toxicology and Pharmacology* 17: 287–91.

Environmental Protection Agency. 2004. *Title 40: Protection of Environment* (Revised July 1, 2004). Environmental Protection Agency, http://www.access.gpo.gov/nara/cfr/waisidx_04/40cfr180_04.html.

Eskenazi, B., A. Bradman, and R. Castorina. 1999. "Exposures of Children to Organophosphate Pesticides and Their Potential Adverse Health Effects." *Environmental Health Perspectives* 107 (Supplement 3): 409–19.

Farah, Jumanah. 1994. *Pesticide Policies in Developing Countries: Do They Encourage Excessive Use?* World Bank Discussion Papers, No. 238. Washington, DC: World Bank.

Farley, T. A., and L. McFarland. 1999. "Aldicarb as a Cause of Food Poisoning—Louisiana, 1998." *Morbidity and Mortality Weekly Report* 48: 269–71.

Feder, Gershon. 1980. "Farm Size, Risk Aversion and the Adoption of New Technology under Uncertainty." *Oxford Economic Papers* 32, no. 2: 263–83.

Feder, Gershon, Richard E. Just, and David Zilberman. 1985. "Adoption of Agricultural Innovations in Developing Countries: A Survey." *Economic Development and Cultural Change* 33, no. 2: 255–98.

Feola, Giuseppe, and Claudia R. Binder. 2010. "Why Don't Pesticide Applicators Protect Themselves? Exploring the Use of Personal Protective Equipment among Colombian Smallholders." *International Journal of Occupational and Environmental Health* 16, no. 1: 11–23.

Fernandez-Cornejo, Jorge, Sharon Jans, and Mark Smith. 1998. "Issues in the Economics of Pesticide Use in Agriculture: A Review of the Empirical Evidence." *Review of Agricultural Economics* 20, no. 2: 462–88.

Fischer, Edward F., and Peter Benson. 2006. *Broccoli and Desire: Global Connections and Maya Struggles in Postwar Guatemala.* Stanford, CA: Stanford University Press.

Flower, Kori B., Jane A. Hoppin, Charles F. Lynch, Aaron Blair, Charles Knott, David L. Shore, and Dale P. Sandler. 2004. "Cancer Risk and Parental Pesticide Application in Children of Agricultural Health Study Participants." *Environmental Health Perspectives* 112: 631–35.

Flynn, Andrew, Terry Marsden, and Sarah Whatmore. 1994. "Retailing, the Food System and the Regulatory State." In *Regulating Agriculture,* edited by P. Lowe, T. Marsden, and S. Whatmore, 90–103. London: David Fulton.

Food and Agriculture Organization. 2003. *Development of a Framework for Good Agricultural Practices.* Committee on Agriculture, 17th Session, March 31–April 4, Rome, Food and Agriculture Organization of the United Nations, http://www.fao.org/docrep/meeting/006/y8704e.htm.

———. 2004. *FAOSTAT Database.* Food and Agriculture Organization, http://faostat3.fao.org/home/index.html.

———. 2008. *Codex Alimentarius: Pesticide Residues in Food.* Food and Agriculture Organization, April 9, http://www.codexalimentarius.net/mrls/pestdes/jsp/pest_q-e.jsp.

Food and Drug Administration. 2013. "FDA Pesticide Program Residue Monitoring: 1993–2009," May 21, http://www.fda.gov/Food/FoodborneIllnessContaminants/Pesticides/UCM2006797.htm.

Food and Drug Administration Pesticide Program. 1988. "Residues in Foods—1987." *Journal of the Association of Official Analytical Chemists* 71, no. 6: 156A–74A.

———. 1989. "Residues in Foods—1988." *Journal of the Association of Official Analytical Chemists* 71, no. 6: 133A–52A.

———. 2004. *Residue Monitoring 2002.* Washington, DC: Food and Drug Administration.

Foreign Agricultural Service. 2005a. *Central American–Dominican Republic–United States–Free Trade Agreement: Overall Agriculture Fact Sheet.* U.S. Department of Agriculture, http://www.fas.usda.gov/info/factsheets/CAFTA/overall021105a.html.

———. 2005b. *U.S. Trade Imports: FATUS Commodity Aggregations.* Washington, DC: U.S. Department of Agriculture, Foreign Agricultural Service.

———. 2010. *CAFTA-DR Free Trade Agreement: Early Assessment of the Agreement.* Washington, DC: U.S. Department of Agriculture.

Fournier L., María Luisa, Fernando Ramírez M., Clemens Ruepert, Seiling Vargas V., and Silvia Echeverría S. 2010. *Diagnóstico sobre contaminación de aguas, suelos y productos hortícolas por el uso de agroquímicos en la microcuenca de las quebradas Plantón y Pacayas en Cartago, Costa Rica.* Heredia, Costa Rica: Instituto Regional de Estudios en Sustancias Tóxicas (IRET), Universidad Nacional.

Freidberg, Susanne. 2003. "Cleaning Up Down South: Supermarkets, Ethical Trade and African Horticulture." *Social & Cultural Geography* 4, no. 1: 27–43.

———. 2004. *French Beans and Food Scares: Culture and Commerce in an Anxious Age.* New York: Oxford University Press.

———. 2009. *Fresh: A Perishable History.* Cambridge, MA: Belknap Press of Harvard University Press.

Friedland, William H. 1984. "Commodity Systems Analysis: An Approach to the Sociology of Agriculture." In *Research in Rural Sociology and Development,* edited by H. W. Schwarzweller, 221–35. Greenwich, CT: JAI Press.

Friedmann, Harriet. 1978. "Simple Commodity Production and Wage Labour in the American Plains." *Journal of Peasant Studies* 6, no. 1: 71–100.

Friedmann, Harriet, and Philip McMichael. 1989. "Agriculture and the State System: The Rise and Decline of National Agricultures, 1870 to the Present." *Sociologia Ruralis* 29, no. 2: 93–117.

Fuchs, Doris A., Agni Kalfagianni, and Maarten Arentsen. 2009. "Retail Power, Private Standards, and Sustainability in the Global Food System." In *Corporate Power in Global Agrifood Governance,* edited by J. Clapp and D. A. Fuchs, 29–59. Cambridge, MA: MIT Press.

Fuentes Madríz, Gerardo. 1972. "Evaluación de los suelos de la zona oeste del cantón de Oreamuno." Unpublished thesis, Ingeniero Agrónomo, Facultad de Agronomía, Universidad de Costa Rica, San José, Costa Rica.

Gaete, Marcelo. 1990. *¿Qué es la Agricultura de Cambio?* San José, Costa Rica: Centro de Estudios y Publicaciones Alfoja.

Galt, Ryan E. 2001. "The Pesticide Paradox: Alternative Agriculture and the Political Ecology of Pesticide Use in Costa Rica." Master's thesis, University of Wisconsin–Madison.

———. 2007. "Regulatory Risk and Farmers' Caution with Pesticides in Costa Rica." *Transactions of the Institute of British Geographers* 32, no. 3: 377–94.

———. 2008a. "Beyond the Circle of Poison: Significant Shifts in the Global Pesticide Complex, 1976–2008." *Global Environmental Change* 18, no. 4: 786–99.

———. 2008b. "Pesticides in Export and Domestic Agriculture: Reconsidering Market Orientation and Pesticide Use in Costa Rica." *Geoforum* 39, no. 3: 1378–92.

———. 2008c. "Toward an Integrated Understanding of Pesticide Use Intensity in Costa Rican Vegetable Farming." *Human Ecology* 36, no. 5: 655–77.

———. 2009. "Overlap of U.S. FDA Residue Tests and Pesticides Used on Imported Vegetables: Empirical Findings and Policy Recommendations." *Food Policy* 34: 468–76.

———. 2010. "Scaling Up Political Ecology: The Case of Illegal Pesticides on Fresh Vegetables Imported into the United States, 1996–2006." *Annals of the Association of American Geographers* 100, no. 2: 327–55.

———. 2011. "Circulating Science, Incompletely Regulating Commodities: Governing from a Distance in Transnational Agro-food Networks." In *Knowing Nature: Conversations at the Intersection of Political Ecology and Science Studies*, edited by M. Goldman, M. D. Turner, and P. Nadasdy, 227–43. Chicago: University of Chicago Press.

———. 2013. "From *Homo Economicus* to Complex Subjectivities: Reconceptualizing Farmers as Pesticide Users." *Antipode* 45, no. 2: 336–56.

Gamboa, William, and Omar Serrano. 2003. "Avances en el establecimiento y manejo de una finca agroecológica: Una experiencia con agricultores del Valle de Ujarrás, Costa Rica." *Foro Latinoamericano*, no. 32: 88–106.

García, Jaime E. 1990. "Residuos de plaguicidas en los alimentos: Aspectos introductores." *Tecnología en Marcha* 10, no. 4: 37–41.

———. 1997. *Introducción a los plaguicidas*. San José, Costa Rica: Editorial Universidad Estatal a Distancia.

———. 1999. "El mito del manejo seguro de los plaguicidas en los países en desarrollo." *Manejo Integrado de Plagas*, no. 52: 25–41.

Garry, Vincent F., Dina Schreinemachers, Mary E. Harkins, and Jack Griffith. 1996. "Pesticide Appliers, Biocides, and Birth Defects in Rural Minnesota." *Environmental Health Perspectives* 104, no. 4: 394–99.

General Accounting Office. 1979. *Better Regulation of Pesticide Exports and Pesticide Residues in Imported Food Is Essential*. Washington, DC: General Accounting Office.

———. 1990. *Five Latin American Countries' Controls over the Registration and Use of Pesticides*. Washington, DC: General Accounting Office.

———. 1992. *Adulterated Imported Foods Are Reaching U.S. Grocery Shelves*. Washington, DC: General Accounting Office.

Gereffi, Gary, and Miguel Korzeniewicz, eds. 1994. *Commodity Chains and Global Capitalism*. Westport, CT: Praeger.

Gockowski, James, and Michel Ndoumbe. 2004. "The Adoption of Intensive Monocrop Horticulture in Southern Cameroon." *Agricultural Economics* 30, no. 3: 195–202.

González, Humberto. 2001. "Las redes transnacionales y las cadenas globales de mercancías: La agricultura de exportación en México." *Amérique Latine Histoire et Mémoire* 2, http://alhim.revues.org/613.

González, Luis Carlos. 1975. *Principales enfermedades de los cultivos de Costa Rica.* San José: Facultad de Agronomía, Universidad de Costa Rica.

Goodman, David, and Michael Redclift. 1981. *From Peasant to Proletarian: Capitalist Development and Agrarian Transitions.* Oxford, UK: Basil Blackwell.

Goodman, David, Bernardo Sorj, and John Wilkinson. 1987. *From Farming to Biotechnology: A Theory of Agro-industrial Development.* New York: Basil Blackwell.

Goodman, David, and Michael Watts, eds. 1997. *Globalising Food: Agrarian Questions and Global Restructuring.* New York: Routledge.

Graham, Frank. 1970. *Since Silent Spring.* Boston: Houghton-Mifflin.

Green, Margaret A., Michael A. Heumann, H. Michael Wehr, Laurence R. Foster, L. Paul Williams Jr., Jacquelyn A. Polder, Clarence L. Morgan, Sheldon L. Wagner, Lee A. Wanke, and James M. Witt. 1987. "An Outbreak of Watermelon-borne Pesticide Toxicity." *American Journal of Public Health* 77, no. 11: 1431–34.

Grossman, Lawrence S. 1992. "Pesticides, Caution, and Experimentation in St. Vincent, Eastern Caribbean." *Human Ecology* 20, no. 3: 315–36.

———. 1998. *The Political Ecology of Bananas: Contract Farming, Peasants, and Agrarian Change in the Eastern Caribbean.* Chapel Hill: University of North Carolina Press.

Groth, Edward, III, Charles M. Benbrook, and Karen Lutz. 2000. *Update: Pesticides in Children's Foods; An Analysis of 1998 USDA PDP Data on Pesticide Residues.* Yonkers, NY: Consumers Union of U.S.

Guan-Soon, Lim, and Ong Seng-Hock. 1987. "Environmental Problems of Pesticide Usage in Malaysian Rice Fields—Perceptions and Future Considerations." In *Management of Pests and Pesticides: Farmers' Perceptions and Practices,* edited by J. Tait and B. Napompeth, 10–21. Boulder, CO: Westview.

Guillette, Elizabeth A., Craig Conard, Fernando Lares, Maria Guadalupe Aguilar, John McLachlan, and Louis J. Guillette Jr. 2006. "Altered Breast Development in Young Girls from an Agricultural Environment." *Environmental Health Perspectives* 114, no. 3: 471–75.

Guillette, Elizabeth A., Maria Mercedes Meza, Maria Guadalupe Aguilar, Alma Delia Soto, and Idalia Enedina Garcia. 1998. "An Anthropological Approach to the Evaluation of Preschool Children Exposed to Pesticides in Mexico." *Environmental Health Perspectives* 106, no. 6: 347–53.

Gupta, P. K. 2004. "Pesticide Exposure—Indian Scene." *Toxicology* 198, nos. 1–3: 83–90.

Guthman, Julie. 2004. *Agrarian Dreams: The Paradox of Organic Farming in California.* Berkeley: University of California Press.

———. 2007. "The Polanyian Way? Voluntary Food Labels as Neoliberal Governance." *Antipode* 39, no. 3: 456–78.

Gutiérrez C., Fernando. 2003. "Intoxicados niños con agroquímico." *La Nación,* October 29, http://www.nacion.com/ln_ee/2003/octubre/29/pais16.html.

Gwynne, Robert N. 1993. "Outward Orientation and Marginal Environments: The Question of Sustainable Development in the Norte Chico, Chile." *Mountain Research and Development* 13, no. 3: 281–93.

———. 1999. "Globalisation, Commodity Chains and Fruit Exporting Regions in Chile." *Tijdschrift voor Economische en Sociale Geografie* 90, no. 2: 211–25.

Gwynne, Robert N., and Jorge Ortiz. 1997. "Export Growth and Development in Poor Rural Regions: A Meso-Scale Analysis of the Upper Limari." *Bulletin of Latin American Research* 16, no. 1: 25–41.

Hamilton, Sarah, Linda Asturias de Barrios, and Brenda Tevalán. 2001. "Gender and Commercial Agriculture in Ecuador and Guatemala." *Culture and Agriculture* 23, no. 3: 1–12.

Hamilton, Sarah, and Edward F. Fischer. 2003. "Non-traditional Agricultural Exports in Highland Guatemala: Understandings of Risk and Perceptions of Change." *Latin American Research Review* 38, no. 3: 82–110.

———. 2005. "Maya Farmers and Export Agriculture in Highland Guatemala: Implications for Development and Labor Relations." *Latin American Perspectives* 32, no. 5: 33–58.

Hardell, Lennart, and Mikael Eriksson. 1999. "A Case-Control Study of Non-Hodgkin Lymphoma and Exposure to Pesticides." *Cancer* 85: 1353–60.

Harrison, Jill. 2008. "Lessons Learned from Pesticide Drift: A Call to Bring Production Agriculture, Farm Labor, and Social Justice Back into Agrifood Research and Activism." *Agriculture and Human Values* 25, no. 2: 163–67.

Harrison, Jill Lindsey. 2006. "'Accidents' and Invisibilities: Scaled Discourse and the Naturalization of Regulatory Neglect in California's Pesticide Drift Conflict." *Political Geography* 25, no. 5: 506–29.

Harvey, David. 1999. *The Limits to Capital.* New York: Verso.

Hausbeck, Mary K., and Kurt H. Lamour. 2004. "*Phytophthora capsici* on Vegetable Crops: Research Progress and Management Challenges." *Plant Disease* 88, no. 12: 1292–1303.

Haverkort, A. J. 1990. "Ecology of Potato Cropping Systems in Relation to Latitude and Altitude." *Agricultural Systems* 32: 251–72.

Hawken, Paul. 1993. *The Ecology of Commerce.* New York: HarperCollins.

Hawkins, Keith. 1989. "Rule and Discretion in Comparative Perspective: The Case of Social Regulation." *Ohio State Law Journal* 50: 663–79.

Hecht, Susanna B., and Alexander Cockburn. 1990. *The Fate of the Forest: Developers, Destroyers and Defenders of the Amazon.* New York: Harper Perennial.

Heong, K. L., M. M. Escalada, and A. A. Lazaro. 1995. "Misuse of Pesticides among Rice Farmers in Philippines." In *Impact of Pesticides on Farmer Health and the Rice Environment,* edited by P. L. Pingali and P. A. Roger, 97–108. Norwell, MA, and Los Baños, Philippines: Kluwer Academic Publishers and IRRI.

Hijmans, Robert J., G. A. Forbes, and T. S. Walker. 2000. "Estimating the Global Severity of Potato Late Blight with GIS-Linked Disease Forecast Models." *Plant Pathology* 49: 697–705.

Hilje, Luko, Luisa E. Castillo M., Lori Ann Thrupp, and Ineke Wesseling H. 1987. *El uso de los plaguicidas en Costa Rica.* San José, Costa Rica: Editorial Universidad Estatal a Distancia.

Hill, Dennis S., and J. M. Waller. 1988. *Pests and Diseases of Tropical Crops,* Vol. 1, *Principles and Methods of Control.* New York: Longman.

Holland, Rob. 1998. "Food Product Liability Insurance." ADC Info #11, Agricultural Development Center, Agricultural Extension Service, University of Tennessee.

Holt-Giménez, Eric. 2006. *Campesino a Campesino: Voices from Latin America's Farmer to Farmer Movement for Sustainable Agriculture.* Oakland, CA: Food First Books.

Hord, Melanie J., and Carmen Rivera. 1998. "Prevalencia y distribución geográfica de los virus PVX, PVY, PVA, PVM, PVS y PLRV en el cultivo de la papa en la Zona Norte de Cartago, Costa Rica." *Agronomía Costarricense* 22, no. 2: 137–43.

Horst, Oscar H. 1987. "Commercialization of Traditional Agriculture in Highland Guatemala and Ecuador." *Revista Geográfica del Instituto Panamericano de Geografía e Historia* 106: 5–18.

———. 1989. "The Persistence of Milpa Agriculture in Highland Guatemala." *Journal of Cultural Geography* 9, no. 2: 13–30.

Horton, Douglas E. 1987. *Potatoes: Production, Marketing, and Programs in Developing Countries.* Boulder, CO: Westview.

Hui, Xu, Qian Yi, Peng Bu-zhuo, Jiang Xiliu, and Hua Xiao-mei. 2003. "Environmental Pesticide Pollution and Its Countermeasures in China." *Ambio* 32, no. 1: 78–80.

Hynes, H. Patricia. 1989. *The Recurring Silent Spring.* New York: Pergamon.

Instituto Geográfico Nacional. 1981a. *Istarú.* San José, Costa Rica: Instituto Geográfico Nacional, Ministerio de Obras Públicas y Transportes.

———. 1981b. *Tapantí.* San José, Costa Rica: Instituto Geográfico Nacional, Ministerio de Obras Públicas y Transportes.

Instituto Regional de Estudios en Sustancias Tóxicas. 1999. *Manual de plaguicidas: Guía para América Central.* Heredia, Costa Rica: Editorial de la Universidad Nacional.

International Program on Chemical Safety. 2002. *Global Assessment of the State-of-the-Science of Endocrine Disruptors.* Geneva: World Health Organization.

Ip, Henrietta Man Hing. 1990. "Chlorinated Pesticides in Foodstuffs in Hong Kong." *Archives of Environmental Contamination and Toxicology* 19, no. 2: 291–96.

Jackson, M. T. 1983. "Potatoes (Papas)." In *Costa Rican Natural History,* edited by D. H. Janzen, 103–5. Chicago: University of Chicago Press.

Jackson, M. T., Luis F. Cartín, and Jorge A. Aguilar 1981. "Uso y manejo de fertilizantes en el cultivo de la papa (*Solanum tuberosum* L.) en Costa Rica." *Agronomía Costarricense* 5, nos. 1–2: 15–19.

Jaffee, Daniel. 2007. *Brewing Justice: Fair Trade Coffee, Sustainability, and Survival.* Berkeley: University of California Press.

Jaffee, Steven. 1993. *Exporting High-Value Food Commodities: Success Stories from Developing Countries.* Washington, DC: World Bank.

———. 1995. "The Many Faces of Success: The Development of Kenyan Horticultural Exports." In *Marketing Africa's High-Value Foods: Comparative Experiences of an Emergent Private Sector,* edited by S. Jaffee and J. Morton, 319–74. Dubuque: Kendall/Hunt.

Jansen, Hans G. P., Aad van Tilburg, John Belt, and Susan Hoekstra. 1996. *Agricultural Marketing in the Atlantic Zone of Costa Rica: A Production, Consumption and Trade Study of Agricultural Commodities Produced by Small and Medium-Scale Farmers.* Turrialba: CATIE.

Jansen, Kees. 2008. "The Unspeakable Ban: The Translation of Global Pesticide Governance into Honduran National Regulation." *World Development* 36, no. 4: 575–89.

Jessop, Bob. 2001. "Regulationist and Autopoieticist Reflections on Polanyi's Account of Market Economies and the Market Society." *New Political Economy* 6, no. 2: 213–32.

Jeyaratnam, J. 1990. "Acute Pesticide Poisoning: A Major Global Health Problem." *World Health Statistics Quarterly* 43: 139–44.

Jiménez París, Ana Lorena. 1996. "Aplicación metodológica para el cálculo de un índice de precios de insumos: Caso de la papa en la zona norte de la provincia de Cartago." Unpublished thesis, Facultad de Agronomía, Universidad de Costa Rica, Ciudad Universitaria Rodrigo Facio, San José, Costa Rica.

Johnson, K., E. A. Olson, and S. Manandhar. 1982. "Environmental Knowledge and Response to Natural Hazards in Mountainous Nepal." *Mountain Research and Development* 2, no. 2: 175–88.

Jorgenson, Andrew K. 2007. "Foreign Direct Investment and Pesticide Use Intensity in Less-Developed Countries: A Quantitative Investigation." *Society & Natural Resources* 20, no. 1: 73–83.

Julian, James W., Glenn H. Sullivan, and Guillermo E. Sánchez. 2000. "Future Market Development Issues Impacting Central America's Nontraditional Agricultural Export Sector: Guatemala Case Study." *American Journal of Agricultural Economics* 82, no. 5: 1177–83.

Jungbluth, Frauke. 1997. "Analysis of Crop Protection Policy in Thailand." *Thailand Development Research Institute Quarterly Review* 12, no. 1: 16–23.

Kannan, K., S. Tanabe, A. Ramesh, A. Subramanian, and R. Tatsukawa. 1992. "Persistent Organochlorine Residues in Foodstuffs from India and Their Implications on Human Dietary Exposure." *Journal of Agricultural and Food Chemistry* 40, no. 3: 518–24.

Kansakar, Vidya Bir Singh, Narendra Raj Khanal, and Motilal Ghimire. 2002. "Use of Insecticides in Nepal." *Landschaftsökologie und Umweltforschung* 38: 90–98.

Keese, James R. 1998. "International NGOs and Land Use Change in a Southern Highland Region of Ecuador." *Human Ecology* 26, no. 3: 451.

Kegley, Susan, Stephan Orme, and Lars Neumeister. 2000. *Hooked on Poison: Pesticide Use in California, 1991–1998*. San Francisco: Pesticide Action Network North America.

Kinkela, David. 2011. *DDT and the American Century: Global Health, Environmental Politics, and the Pesticide That Changed the World*. Chapel Hill: University of North Carolina Press.

Kirkby, Anne V. 1974. "Individual and Community Responses to Rainfall Variability in Oaxaca, Mexico." In *Natural Hazards: Local, National, Global*, edited by G. F. White, 119–28. New York: Oxford University Press.

Kloppenburg, Jack, Jr., and Beth Burrows. 2001. "Biotechnology to the Rescue? Ten Reasons Why Biotechnology Is Incompatible with Sustainable Agriculture." In *Redesigning Life? The Worldwide Challenge to Genetic Engineering*, edited by B. Tokar, 103–10. New York: Zed Books.

Kloppenburg, Jack Ralph. 1988. *First the Seed: The Political Economy of Plant Biotechnology, 1492–2000*. New York: Cambridge University Press.

Knight, Andrew, and Rex Warland. 2004. "The Relationship between Sociodemographics and Concern about Food Safety Issues." *Journal of Consumer Affairs* 38, no. 1: 107–20.

Konefal, Jason, Michael Mascarenhas, and Maki Hatanaka. 2005. "Governance in the Global Agro-Food System: Backlighting the Role of Transnational Supermarket Chains." *Agriculture and Human Values* 22, no. 3: 291–302.

Kopper, G. 2002. "Food Safety Perspectives in Costa Rica: Export and National Markets for Fresh Produce." In *Food Safety Management in Developing Countries: Proceedings of the International Workshop, CIRAD-FAO, 11–13 December 2000*, edited by E. Hanak, E. Boutrif, P. Fabre, and M. Pineiro, 1–4. Montpellier, France: CIRAD-FAO.

Kremen, Claire, Alastair Iles, and Christopher Bacon. 2012. "Diversified Farming Systems: An Agroecological, Systems-based Alternative to Modern Industrial Agriculture." *Ecology and Society* 17, no. 4: Article 44, http://www.ecologyandsociety.org/vol17/iss4/art44/.

Krimsky, Sheldon. 2000. "Environmental Endocrine Hypothesis and Public Policy." In *Illness and the Environment: A Reader in Contested Medicine*, edited by S. Kroll-Smith, P. Brown, and V. J. Gunter, 95–107. New York: New York University Press.

Kuepper, George. 2003. *Downy Mildew Control in Cucurbits*. Fayetteville, AR: Appropriate Technology Transfer for Rural Areas.

Kuhlken, Robert. 1999. "Settin' the Woods on Fire: Rural Incendarism as Protest." *Geographical Review* 89, no. 3: 343–63.

La Gaceta. 1997. Límites máximos de residuos de plaguicidas en vegetales, Decreto Ejecutivo 26031-MEIC-MAG-S, No. 103. San José, Costa Rica.

Lappé, Frances Moore. 1991. *Diet for a Small Planet*. 20th anniversary ed. New York: Ballantine Books.

Lappé, Frances Moore, Joseph Collins, Peter Rosset, and Luis Esparza. 1999. *World Hunger: Twelve Myths.* New York: Grove.

Le Heron, Richard, and Michael Roche. 1999. "Rapid Reregulation, Agricultural Restructuring and the Reimagining of Agriculture in New Zealand." *Rural Sociology* 64, no. 2: 203–18.

Lehman, Hugh. 1993. "Values, Ethics, and the Use of Synthetic Pesticides in Agriculture." In *The Pesticide Question: Environment, Economics, and Ethics,* edited by D. Pimentel and H. Lehman, 347–79. New York: Chapman and Hall.

Leveridge, Yamillette R. 1998. "Pesticide Poisoning in Costa Rica during 1996." *Veterinary and Human Toxicology* 40, no. 1: 42–44.

Levins, Richard A. 2000. *Willard Cochrane and the American Family Farm.* Lincoln: University of Nebraska Press.

Lewis, Martin W. 1989. "Commercialization and Community Life: The Geography of Market Exchange in a Small-Scale Philippine Society." *Annals of the Association of American Geographers* 79, no. 3: 390–410.

———. 1992. *Wagering the Land: Ritual, Capital, and Environmental Degradation in the Cordillera of Northern Luzon, 1900–1986.* Berkeley: University of California Press.

Lewontin, R. C. 2000. "The Maturing of Capitalist Agriculture: Farmer as Proletarian." In *Hungry for Profit: The Agribusiness Threat to Farmers, Food, and the Environment,* edited by F. Magdoff, J. B. Foster, and F. H. Buttel, 93–106. New York: Monthly Review Press.

Lewontin, Richard C., and Richard Levins. 2007. *Biology under the Influence: Dialectical Essays on Ecology, Agriculture, and Health.* New York: Monthly Review Press.

Li, Y. F., J. Struger, D. Waite, and J. Ma. 2004. "Gridded Canadian Lindane Usage Inventories with 1/6° x 1/4° Latitude and Longitude Resolution." *Atmospheric Environment* 38: 1117–21.

Lira Saade, Rafael, and S. Montes Hernández. 1994. "*Cucurbita Pepo.*" In *Neglected Crops: 1492 from a Different Perspective,* edited by J. E. Hernándo Bermejo and J. León, 63–77. Rome: Food and Agriculture Organization.

Little, Peter D. 1994. "Contract Farming and the Development Question." In *Living under Contract: Contract Farming and Agrarian Transformation in Sub-Saharan Africa,* edited by P. D. Little and M. Watts, 216–47. Madison: University of Wisconsin Press.

Lizano, E. 1994. "The Impact of Policy Reform on Food Consumption in Costa Rica." PhD dissertation, Fletcher School of Law and Diplomacy, Tufts University, Medford, MA.

Lockie, Stewart. 2006. "Networks of Agri-environmental Action: Temporality, Spatiality and Identity in Agricultural Environments." *Sociologia Ruralis* 46, no. 1: 22–39.

Lomborg, Bjørn. 2001. *The Skeptical Environmentalist: Measuring the Real State of the World.* New York: Cambridge University Press.

López-García, Rebeca. 2003. "United States Import/Export Regulations and Certification." In *Food Safety Handbook,* edited by R. H. Schmidt and G. E. Rodrick, 741–58. Hoboken, NJ: Wiley.

Lowe, Philip, Terry Marsden, and Sarah Whatmore. 1994. "Changing Regulatory Orders: The Analysis of the Economic Governance of Agriculture." In *Regulating Agriculture,* edited by P. Lowe, T. Marsden, and S. Whatmore, 1–30. London: David Fulton Publishers.

Lu, Chensheng, Kathryn Toepel, Rene Irish, Richard A. Fenske, Dana B. Barr, and Roberto Bravo. 2006. "Organic Diets Significantly Lower Children's Dietary Exposure to Organophosphorus Pesticides." *Environmental Health Perspectives* 114, no. 2: 260–63.

Lyson, Thomas A. 2004. *Civic Agriculture: Reconnecting Farm, Food, and Community.* Medford, MA: Tufts University Press.

Mann, Susan, and James M. Dickinson. 1978. "Obstacles to the Development of a Capitalist Agriculture." *Journal of Peasant Studies* 5, no. 4: 466–81.

Mannon, Susan E. 1998. "Negotiating Risk in Agricultural Production: Case Studies from the Non-traditional Agricultural Export Sectors of Kenya and Costa Rica." Master's thesis, University of Wisconsin–Madison.

———. 2005. "Risk Takers, Risk Makers: Small Farmers and Non-traditional Agroexports in Kenya and Costa Rica." *Human Organization* 64, no. 1: 16–27.

Marsden, Terry. 1988. "Exploring Political Economy Approaches in Agriculture." *Area* 20: 315–21.

Marsden, Terry, Andrew Flynn, and Michelle Harrison. 1999. *Consuming Interests: The Social Provision of Foods.* London: UCL Press.

Marsden, Terry, Richard Munton, Neil Ward, and Sarah Whatmore. 1996. "Agricultural Geography and the Political Economy Approach: A Review." *Economic Geography,* Special issue, *The New Rural Geography* 72, no. 4: 361–75.

Marx, Karl. 1990. *Capital: A Critique of Political Economy,* Vol. 1. New York: Penguin in association with New Left Review.

———. 1991. *Capital: A Critique of Political Economy,* Vol. 3. New York: Penguin in association with New Left Review.

Matson, P. A., W. J. Parton, A. G. Power, and M. J. Swift. 1997. "Agricultural Intensification and Ecosystem Properties." *Science* 277, no. 5325: 504–9.

Matute Ch., Ronald. 1991. "MAG suspendió uso de pesticida en banano." *La Nación,* August 6, 4A.

McCarthy, James. 2005. "First World Political Ecology: Directions and Challenges." *Environment and Planning A* 37, no. 6: 953–58.

McMichael, Philip, ed. 1994. *The Global Restructuring of Agro-food Systems.* Ithaca, NY: Cornell University Press.

———, ed. 1995. *Food and Agrarian Orders in the World-Economy.* Westport, CT: Greenwood.

———. 2009. "A Food Regime Genealogy." *Journal of Peasant Studies* 36, no. 1: 139–69.

Medina, Charito P. 1987. "Pest Control Practices and Pesticide Perceptions of Vegetable Farmers in Loo Valley, Benguet, Philippines." In *Management of Pests and Pesticides: Farmers' Perceptions and Practices,* edited by J. Tait and B. Napompeth, 150–57. Boulder, CO: Westview.

Melo, Christian J., and Steven A. Wolf. 2005. "Empirical Assessment of Eco-Certification: The Case of Ecuadorian Bananas." *Organization and Environment* 18, no. 3: 287–317.

Meneses, R. 1990. "Monitoreo de áfidos y su relación con el programa de semilla de papa en Costa Rica." *Manejo Integrado de Plagas,* no. 15: 45–52.

Midmore, D. J., H. G. P. Jansen, R. G. Dumsday, A. A. Azmi, D. D. Poudel, S. Valasayya, J. Huang, M. M. Radzali, N. Fuad, A. B. Samah, A. R. Syed, and A. Nazlin. 1996. *Technical and Economic Aspects of Sustainable Production Practices among Vegetable Growers in the Cameron Highlands, Malaysia.* Taipei, Taiwan: Asian Vegetable Research and Development Center.

Mo, Claudette Lee. 2001. "Environmental Impact of Leatherleaf Fern Farms in Costa Rica." PhD dissertation, University of Wisconsin–Madison.

Mohanty, B. B. 2005. "'We Are Like the Living Dead': Farmer Suicides in Maharashtra, Western India." *Journal of Peasant Studies* 32, no. 2: 243–76.

Molina Umaña, Mauro. 1961. "Evaluación de producción y resistencia a *P. infestans* de 5 clones para nombramiento de variedades." Unpublished thesis, Ingeniero Agrónomo, Facultad de Agronomía, Universidad de Costa Rica, San José, Costa Rica.

Monge, P., C. Wesseling, J. Guardado, I. Lundberg, A. Ahlbom, K. P. Cantor, E. Weiderpass, and T. Partanen. 2007. "Parental Occupational Exposure to Pesticides and the Risk of Childhood Leukemia in Costa Rica." *Scandinavian Journal of Work, Environment & Health* 33, no. 4: 293–303.

Monge V., Luis Alberto, Elizabeth Carazo, Victoria Cartín, and Jorge Lobo. 1996. *Informe final del proyecto de resistencia a los insecticidas de Plutella xylostella (palomilla de las cruciferas) en Costa Rica.* Cartago: Instituto Tecnológico de Costa Rica.

Montaldo, Alvaro. 1984. *Cultivo y mejoramiento de la papa.* San José, Costa Rica: Instituto Interamericano de Cooperación para la Agricultura.

Mooney, Patrick H. 1988. *My Own Boss? Class, Rationality, and the Family Farm.* Boulder, CO: Westview.

Moore, Jason W. 2008. "Ecological Crises and the Agrarian Question in World-Historical Perspective." *Monthly Review* 60, no. 6: 54–63.

Morales, Helda, and Ivette Perfecto. 2000. "Traditional Knowledge and Pest Management in the Guatemalan Highlands." *Agriculture and Human Values* 17, no. 1: 49–63.

Morrison, Paul Cross. 1955. "Land Utilization, Cartago to Turrialba, Costa Rica." *Papers of the Michigan Academy of Science, Arts, and Letters* 40: 205–16.

Moses, Marion. 1993. "Farmworkers and Pesticides." In *Confronting Environmental Racism: Voices from the Grassroots,* edited by R. D. Bullard, 161–78. Boston: South End.

Mukherjee, Irani. 2003. "Pesticides Residues in Vegetables in and around Delhi." *Environmental Monitoring and Assessment* 86, no. 3: 265–71.

Murdoch, Jonathan, Terry Marsden, and Jo Banks. 2000. "Quality, Nature, and Embeddedness: Some Theoretical Considerations in the Context of the Food Sector." *Economic Geography* 76, no. 2: 107–25.

Murray, Douglas L. 1991. "Export Agriculture, Ecological Disruption, and Social Inequity: Some Effects of Pesticides in Southern Honduras." *Agriculture and Human Values* 8, no. 4: 19–29.

———. 1994. *Cultivating Crisis: The Human Cost of Pesticides in Latin America.* Austin: University of Texas Press.

Murray, Douglas L., and Polly Hoppin. 1992. "Recurring Contradictions in Agrarian Development: Pesticide Problems in Caribbean Basin Nontraditional Agriculture." *World Development* 20, no. 4: 597–608.

Murray, Douglas L., and Peter Leigh Taylor. 2000. "Claim No Easy Victories: Evaluating the Pesticide Industry's Global Safe Use Campaign." *World Development* 28, no. 10: 1735–49.

Murray, Warwick E. 1997. "Competitive Global Fruit Export Markets: Marketing Intermediaries and Impacts on Small-Scale Growers in Chile." *Bulletin of Latin American Research* 16, no. 1: 43–55.

———. 1998. "The Globalisation of Fruit, Neo-liberalism and the Question of Sustainability: Lessons from Chile." *European Journal of Development Research* 10, no. 1: 201–27.

Mutersbaugh, Tad. 2002. "The Number Is the Beast: A Political Economy of Organic-Coffee Certification and Producer Unionism." *Environment and Planning A* 34, no. 7: 1165–84.

Nash, Linda Lorraine. 2006. *Inescapable Ecologies: A History of Environment, Disease, and Knowledge.* Berkeley: University of California Press.

National Research Council. 1983. *Risk Assessment in the Federal Government: Managing the Process.* Washington, DC: National Academy Press.

———. 1987. *Regulating Pesticides in Food: The Delaney Paradox.* Washington, DC: National Academy Press.

Nauen, Ralf, and Thomas Bretschneider. 2002. "New Modes of Action of Insecticides." *Pesticide Outlook* 13: 241–45.

Navarro, Patricia, and Marvin Barquero. 1990. "Nadie controla residuos en hortalizas: Sólo un laboratorio analiza alimentos, distribuidores dicen que cumplen ley." *La Nación,* November 14, 5A.

Nestle, Marion. 2003. *Safe Food: Bacteria, Biotechnology, and Bioterrorism.* Berkeley: University of California Press.

Nieto Z., Oscar. 2001. *Fichas técnicas de plaguicidas a prohibir o restringir incluidos en el Acuerdo No. 9 de la XVI Reunión del Sector Salud de Centroamérica y República Dominicana (RESSCAD).* San José, Costa Rica: Organización Panamericana de la Salud/Organización Mundial de la Salud (OPS/OMS).

O'Brien, Mary. 1988. "Quantitative Risk Analysis: Overused, Under-examined." *Journal of Pesticide Reform* 8, no. 1: 7–12.

———. 2000. *Making Better Environmental Decisions: An Alternative to Risk Assessment.* Cambridge, MA: MIT Press.

O'Connor, James. 1993. "Is Sustainable Capitalism Possible?" In *Food for the Future: Conditions and Contradictions of Sustainability,* edited by P. Allen, 125–37. New York: Wiley.

Okello, Julius, and Ruth Okello. 2010. "Do EU Pesticide Standards Promote Environmentally-Friendly Production of Fresh Export Vegetables in Developing Countries? The Evidence from Kenyan Green Bean Industry." *Environment, Development and Sustainability* 12, no. 3: 341–55.

Okello, Julius J., and Scott M. Swinton. 2007. "Compliance with International Food Safety Standards in Kenya's Green Bean Industry: Comparison of a Small- and a Large-Scale Farm Producing for Export." *Review of Agricultural Economics* 29, no. 2: 269–85.

Opondo, Mary Magdalene. 2000. "The Socio-economic and Ecological Impacts of the Agro-industrial Food Chain on the Rural Economy in Kenya." *Ambio* 29, no. 1: 35–41.

Orme, S., and S. Kegley. 2004. "PAN Pesticides Database—Chemicals," Pesticide Action Network (PAN) Pesticide Database, http://www.pesticideinfo.org.

Ortíz Gutiérrez, Javier. 1996. "Agroquímicos y cáncer gástrico." *La Prensa Libre,* December 4, 13.

Osburn, Susan. 2001. *Do Pesticides Cause Lymphoma?* Chevy Chase, MD: Lymphoma Foundation of America.

Ostrom, Elinor, and Roy Gardner. 1993. "Coping with Asymmetries in the Commons: Self-Governing Irrigation Systems Can Work." *Journal of Economic Perspectives* 7, no. 4: 93–112.

Páez, Oswaldo, Luis Gómez, Arturo Brenes, and Roberto Valverde. 2001. "Resistencia de aislamientos de *Phytophthora infestans* al metalaxyl en el cultivo de la papa en Costa Rica." *Agronomía Costarricense* 25, no. 1: 33–44.

Painter, Michael. 1995. "Upland-Lowland Production Linkages and Land Degradation in Bolivia." In *The Social Causes of Environmental Destruction in Latin America,* edited by M. Painter and W. H. Durham, 133–68. Ann Arbor: University of Michigan Press.

Patel, Raj. 2010. *The Value of Nothing: How to Reshape Market Society and Redefine Democracy.* New York: Picador.

Patel, Raj, and Philip McMichael. 2009. "A Political Economy of the Food Riot." *Review* 32, no. 1: 9–35.

Patton, Michael Quinn. 2002. *Qualitative Research and Evaluation Methods.* 3rd ed. Thousand Oaks, CA: Sage.

Paus, Eva A., ed. 1988. *Struggle against Dependence: Nontraditional Export Growth in Central America and the Caribbean.* Boulder, CO: Westview.

Perez, Carlos J., and Anthony M. Shelton. 1997. "Resistance of *Plutella xylostella* (Lepidoptera: Plutellidae) to *Bacillus thuringiensis* Berliner in Central America." *Insecticide Resistance and Resistance Management* 90, no. 1: 87–93.

Perfecto, Ivette, John H. Vandermeer, and Angus Lindsay Wright. 2009. *Nature's Matrix: Linking Agriculture, Conservation and Food Sovereignty.* Sterling, VA: Earthscan.

Perkins, John H. 1990. "The Rockefeller Foundation and the Green Revolution, 1941–1956." *Agriculture and Human Values* 7, nos. 3–4: 6–18.

Picado Rojas, José Luis, and Francisco Ramírez Matamoros. 1998. *Guía de agroquímicos.* San José, Costa Rica: Desarrollo y Registro de Agroquímicos S. A. and Agrocontinental S. A.

Pimentel, David. 1995. "Amounts of Pesticides Reaching Target Pests: Environmental Impacts and Ethics." *Journal of Agricultural and Environmental Ethics* 8, no. 1: 17–29.

———. 2009. "Environmental and Economic Costs of the Application of Pesticides Primarily in the United States." In *Integrated Pest Management: Innovation-Development Process,* edited by R. Peshin and A. K. Dhawan, 89–111. Dordrecht: Springer Netherlands.

Pimentel, David, H. Acquay, M. Biltonen, P. Rice, M. Silva, J. Nelson, V. Lipner, S. Giordano, A. Horowitz, and M. D'Amore. 1992. "Environmental and Economic Costs of Pesticide Use." *BioScience* 42, no. 10: 750–60.

Pimentel, David, and Anthony Greiner. 1997. "Environmental and Socio-economic Costs of Pesticide Use." In *Techniques for Reducing Pesticide Use,* edited by D. Pimentel, 51–78. New York: Wiley.

Pimentel, David, Colleen Kirby, and Anoop Shroff. 1993. "The Relationship between 'Cosmetic Standards' for Foods and Pesticide Use." In *The Pesticide Question: Environment, Economics, and Ethics,* edited by D. Pimentel and H. Lehman, 85–105. New York: Chapman and Hall.

Pimentel, David, and Hugh Lehman, eds. 1993. *The Pesticide Question: Environment, Economics, and Ethics.* New York: Chapman and Hall.

Pineda Cabrales, Marcos A. 1973. "Algunos aspectos sobre el cultivo del chayote (*Sechium edule* SW)." Unpublished thesis, Departamento Fitotecnia, Facultad de Agronomía, Universidad de Costa Rica, San José.

Polanyi, Karl. 1957. *The Great Transformation.* 1st Beacon paperback ed. Boston: Beacon.

Pollan, Michael. 2006. *The Omnivore's Dilemma: A Natural History of Four Meals.* New York: Penguin.

Poppendieck, Janet. 1999. *Sweet Charity? Emergency Food and the End of Entitlement.* New York: Penguin.

Porter, Warren P., James W. Jaeger, and Ian H. Carlson. 1999. "Endocrine, Immune, and Behavioral Effects of Aldicarb (Carbamate), Atrazine (Triazine) and Nitrate (Fertilizer) Mixtures at Groundwater Concentrations." *Toxicology and Industrial Health* 15, nos. 1–2: 133–50.

Povolny, D. 1973. "*Scrobipalpopsis solanivora* [*Tecia solanivora*] sp.: A New Pest of Potato (*Solanum tuberosum*) from Central America." *Acta Universitatis Agriculturae, Facultaa Agronomica* 21, no. 1: 133–46.

Probst, Kirsten, L. Pülschen, J. Sauerborn, and C. P. W. Zebitz. 1999. "Influencia de varios regímenes de uso de plaguicidas sobre la entomofauna de tomate en las tierras altas de Ecuador." *Manejo Integrado de Plagas*, no. 54: 53–62.

Programa Integral de Mercadeo Agropecuario. 2010. *Tendencias del consumo de frutas, hortalizas, pescado y mariscos en las familias de Costa Rica*. Heredia: Programa Integral de Mercadeo Agropecuario.

Programa Regional Cooperativo de Papa (PRECODEPA). 1982. *Programa nacional de producción de semilla de papa, 1982–1986*. San José, Costa Rica: Ministerio de Agricultura y Ganadería.

Pujara, Dammar Singh, and Narendra Raj Khanal. 2002. "Use of Pesticides in Jaishidihi Sub-catchment, Jhikhu Khola Watershed, Middle Mountain in Nepal." *Landschaftsökologie und Umweltforschung* 38: 168–77.

Pulido, Laura. 1996. *Environmentalism and Economic Justice: Two Chicano Struggles in the Southwest*. Tucson: University of Arizona Press.

Pulido, Laura, and Devon Pena. 1998. "Environmentalism and Positionality: The Early Pesticide Campaign of the United Farm Workers' Organizing Committee, 1965–1971." *Race, Gender & Class* 6, no. 1: 33–50.

Ramírez Aguilar, Carlos Roberto. 1994. "El cultivo de la papa." In *Atlas agropecuario de Costa Rica*, edited by G. Cortés Enríquez, 419–27. San José, Costa Rica: Editorial Universidad Estatal a Distancia.

Raub, Erin. 2008. "Costa Rica Cultivating Fungus-Resistant Potato Varieties." *CostaRicaPages: Travel & Business News*, May 19.

Rauh, Virginia, Srikesh Arunajadai, Megan Horton, Frederica Perera, Lori Hoepner, Dana B. Barr, and Robin Whyatt. 2011. "Seven-Year Neurodevelopmental Scores and Prenatal Exposure to Chlorpyrifos, a Common Agricultural Pesticide." *Environmental Health Perspectives* 119, no. 8: 1196–1201.

Raynolds, Laura T. 1994. "The Restructuring of Third World Agro-exports: Changing Production in the Dominican Republic." In *The Global Restructuring of Agro-food Systems*, edited by P. McMichael, 214–37. Ithaca, NY: Cornell University Press.

———. 1997. "Restructuring National Agriculture, Agro-food Trade, and Agrarian Livelihoods in the Caribbean." In *Globalising Food: Agrarian Questions and Global Restructuring*, edited by D. Goodman and M. Watts, 119–32. New York: Routledge.

———. 2000. "Negotiating Contract Farming in the Dominican Republic." *Human Organization* 59, no. 4: 441–51.

Reardon, Thomas, and Christopher B. Barrett. 2000. "Agroindustrialization, Globalization, and International Development: An Overview of Issues, Patterns, and Determinants." *Agricultural Economics* 23, no. 3: 195–205.

Reardon, Thomas, J. M. Codron, L. Busch, J. Bingen, and C. Harris. 2001. "Global Change in Agrifood Grades and Standards: Agribusiness Strategic Responses in

Developing Countries." *International Food and Agribusiness Management Review* 2, nos. 3–4: 421–25.

Reardon, Thomas, and Stephen A. Vosti. 1995. "Links between Rural Poverty and the Environment in Developing Countries: Asset Categories and Investment Poverty." *World Development* 23, no. 9: 1495–506.

Repetto, Robert, and Sanjay S. Baliga. 1996. *Pesticides and the Immune System: The Public Health Risks*. Baltimore: World Resources Institute.

Ricardo, David. 1971. *On the Principles of Political Economy and Taxation*. Harmondsworth, UK: Penguin.

Richard, Sophie, Safa Moslemi, Herbert Sipahutar, Nora Benachour, and Gilles-Eric Seralini. 2005. "Differential Effects of Glyphosate and Roundup on Human Placental Cells and Aromatase." *Environmental Health Perspectives* 113, no. 6: 716–20.

Ripley, Brian D., Gwen M. Ritcey, C. Ronald Harris, Mary Anne Denommè, and Pamela D. Brown. 2001. "Pyrethroid Insecticide Residues on Vegetable Crops." *Pest Management Science* 57, no. 8: 683–87.

Robbins, Paul. 1998. "Nomadization in Rajasthan, India: Migration, Institutions, and Economy." *Human Ecology* 26, no. 1: 87–112.

———. 2007. *Lawn People: How Grasses, Weeds, and Chemicals Make Us Who We Are*. Philadelphia: Temple University Press.

———. 2012. *Political Ecology: A Critical Introduction*. 2nd ed. Malden, MA: Wiley-Blackwell.

Robbins, Paul, and Julie T. Sharp. 2003. "The Lawn-Chemical Economy and Its Discontents." *Antipode* 35, no. 5: 955–79.

Roberts, Darren M., Ayanthi Karunarathna, Nick A. Buckley, Gamini Manuweera, M. H. Rezvi Sheriff, and Michael Eddleston. 2003. "Influence of Pesticide Regulation on Acute Poisoning Deaths in Sri Lanka." *Bulletin of the World Health Organization* 81, no. 11: 789–98.

Robertson, Morgan. 2004. "The Neoliberalization of Ecosystem Services: Wetland Mitigation Banking and Problems in Environmental Governance." *Geoforum* 35, no. 3: 361–73.

Rodríguez, Carlos L. 1997. "La investigación en *Liriomyza huidobrensis* en el cultivo de papa en Cartago, Costa Rica." *Manejo Integrado de Plagas*, no. 46: 1–8.

Rodríguez, César, Lore Lang, Amy Wang, Karlheinz Altendorf, Fernando García, and André Lipski. 2006. "Lettuce for Human Consumption Collected in Costa Rica Contains Complex Communities of Culturable Oxytetracycline- and Gentamicin-Resistant Bacteria." *Applied and Environmental Microbiology* 72, no. 9: 5870–76.

Rodríguez Navarro, Luis Diego. 1983. "Determinación de residuos del insecticida methamidophos (o,s-dimetil-fosforamidotioato) en lechuga." Unpublished thesis, Ingeniero Agrónomo, Facultad de Agronomía, Universidad de Costa Rica, San José, Costa Rica.

Rodríguez Solano, José Armando. 1994. *Programa de monitoreo de residuos de plaguicidas en muestras de vegetales destinadas al consumo nacional*. San José, Costa

Rica: Laboratorio de Residuos de Plaguicidas, Departamento de Abonos y Plaguicidas, Dirección General de Sanidad Vegetal, M.A.G.

Rojas, José Enrique. 2004. "61 niños intoxicados con plaguicida." *La Nación,* May 15, http://www.nacion.com/ln_ee/2004/mayo/15/pais4.html.

Rosset, Peter M. 1991. "Sustainability, Economies of Scale, and Social Instability: Achilles Heel of Non-traditional Export Agriculture?" *Agriculture and Human Values* 8, no. 4: 30–37.

Russell, Edmund. 2001. *War and Nature: Fighting Humans and Insects with Chemicals from World War I to Silent Spring.* New York: Cambridge University Press.

Saborío Mora, Mario. 1994. "Hortalizas." In *Atlas agropecuario de Costa Rica,* edited by G. Cortés Enríquez, 397–418. San José, Costa Rica: Editorial Universidad Estatal a Distancia.

Sáenz Maroto, Alberto. 1955. *La papa: Curso sinóptico para cultivo de la papa.* San José: Universidad de Costa Rica, Departamento de Publicaciones.

Sánchez Garita, Vera, Richard C. Shattock, and Elkin Bustamante. 2000. "Caracterización de aislamientos de *Phytophthora infestans* nativos de Costa Rica." *Manejo Integrado de Plagas,* no. 55: 36–42.

Sánchez Saldaña, K., and P. Betanzos Ocampo. 2006. "Aspectos socioeconómicos y culturales en el uso de agroquímicos y plaguicidas en los Altos de Morelos, México." *Revibec: Revista Iberoamericana de Economía Ecológica* 3: 33–47.

Sarmah, Ajit K., Karin Müller, and Riaz Ahmad. 2004. "Fate and Behaviour of Pesticides in the Agroecosystem: A Review with a New Zealand Perspective." *Australian Journal of Soil Research* 42, no. 2: 125–54.

Sauer, Carl O. 1969. *Agricultural Origins and Dispersals: The Domestication of Animals and Foodstuffs.* 2nd ed. Cambridge, MA: MIT Press.

Sayer, Andrew. 1992. *Method in Social Science: A Realist Approach.* 2nd ed. London: Routledge.

Schwab, Arnold. 1995. "Pesticides—Aid to Agriculture or Poison—and the Conflicting Role of the Development Worker." In *Pesticides in Tropical Agriculture: Hazards and Alternatives,* edited by R. Altenburger, 107–47. Weikersheim, Germany: Margraf Verlag.

Scott, James C. 1985. *Weapons of the Weak: Everyday Forms of Peasant Resistance.* New Haven, CT: Yale University Press.

———. 1998. *Seeing Like a State: How Certain Schemes to Improve the Human Condition Have Failed.* New Haven, CT: Yale University Press.

Secretaría Ejecutiva de Planificación Sectorial Agropecuaria. 2004. *Costa Rica: Boletín estadístico agropecuario nacional no. 15.* San José, Costa Rica: SEPSA.

———. 2011. *Costa Rica: Boletín estadístico agropecuario nacional no. 21.* San José, Costa Rica: SEPSA.

Segura Coto, Gregorio. 1983. "Análisis económico de sistemas de aplicación de plaguicidas en papa (Solanum tuberosum) en Provincia de Cartago." Unpublished thesis, Economía Agrícola, Facultad de Agronomía, Universidad de Costa Rica, San José, Costa Rica.

Seligson, Mitchell A. 1980. *Peasants of Costa Rica and the Development of Agrarian Capitalism.* Madison: University of Wisconsin Press.

Sherwood, Stephen. 2009. "Learning from Carchi: Agricultural Modernisation and the Production of Decline." PhD dissertation, CERES Graduate Research School for Resource Studies for Development, Wageningen University.

Shiva, Vandana. 1991. *The Violence of the Green Revolution: Third World Agriculture, Ecology, and Politics.* Atlantic Highlands, NJ: Zed Books.

Shrader-Frechette, K. S. 1985. *Risk Analysis and Scientific Method: Methodological and Ethical Problems with Evaluating Societal Hazards.* Boston: D. Reidel Publishing Company.

Sinclair, Upton. 1985. *The Jungle.* New York: Penguin.

Slovic, Paul. 1987. "Perception of Risk." *Science* 236, no. 4799: 280–85.

Smith, Neil. 1984. *Uneven Development: Nature, Capital, and the Production of Space.* New York: Blackwell.

Steingraber, Sandra. 1998. *Living Downstream: A Scientist's Personal Investigation of Cancer and the Environment.* New York: Vintage Books.

Stern, Kingsley Rowland. 1997. *Introductory Plant Biology.* 7th ed. Dubuque, IA: Wm. C. Brown Publishers.

Stewart, Hayden, Noel Blisard, and Dean Jolliffe. 2003. "Do Income Constraints Inhibit Spending on Fruits and Vegetables among Low-Income Households?" *Journal of Agricultural and Resource Economics* 28, no. 3: 465–80.

Stewart, Sarah. 1996. "The Price of a Perfect Flower: Environmental Destruction and Health Hazards in the Colombian Flower Industry." In *Green Guerrillas: Environmental Conflicts and Initiatives in Latin America and the Caribbean,* edited by H. Collinson, 132–39. London: Latin American Bureau.

Stonich, Susan C. 1992. "Struggling with Honduran Poverty: The Environmental Consequences of Natural Resource-Based Development and Rural Transformations." *World Development* 20, no. 3: 385–99.

———. 1993. *I Am Destroying the Land! The Political Ecology of Poverty and Environmental Destruction in Honduras.* Boulder, CO: Westview.

———. 1995. "Development, Rural Impoverishment, and Environmental Destruction in Honduras." In *The Social Causes of Environmental Destruction in Latin America,* edited by M. Painter and W. H. Durham, 63–100. Ann Arbor: University of Michigan Press.

Stonich, Susan C., Douglas L. Murray, and Peter R. Rosset. 1994. "Enduring Crises: The Human and Environmental Consequences of Nontraditional Export Growth in Central America." *Economic Anthropology* 15: 239–74.

Sullivan, Glenn H., Guillermo E. Sánchez, Stephen C. Weller, and C. Richard Edwards. 1999. "Sustainable Development in Central America's Non-traditional Export Crops Sector through Adoption of Integrated Pest Management Practices: Guatemalan Case Study." *Sustainable Development International* 1: 123–26.

Surgeoner, G. A., and W. Roberts. 1993. "Reducing Pesticide Use by 50% in the Province of Ontario: Challenges and Progress." In *The Pesticide Question: Environment,*

Economics, and Ethics, edited by D. Pimentel and H. Lehman, 206–22. New York: Chapman and Hall.

Swyngedouw, Erik. 1999. "Modernity and Hybridity: Nature, *Regeneracionismo,* and the Production of the Spanish Waterscape, 1890–1930." *Annals of the Association of American Geographers* 89, no. 3: 443–65.

Thiers, Paul. 1997. "Successful Pesticide Reduction Policies: Learning from Indonesia." *Society & Natural Resources* 10: 319–28.

Thrupp, Lori Ann. 1988. "Pesticides and Policies: Approaches to Pest-Control Dilemmas in Nicaragua and Costa Rica." *Latin American Perspectives* 15, no. 4: 37–70.

———. 1990a. "The Fallacy of Exporting Risk Analysis to Developing Countries." *Journal of Pesticide Reform* 10, no. 1: 23–25.

———. 1990b. "Inappropriate Incentives for Pesticide Use: Agricultural Credit Requirements in Developing Countries." *Agriculture and Human Values* 7, nos. 3–4: 62–69.

———. 1991a. "Long-term Losses from Accumulation of Pesticide Residues: A Case of Persistent Copper Toxicity in Soils of Costa Rica." *Geoforum* 22, no. 1: 1–15.

———. 1991b. "Sterilization of Workers from Pesticide Exposure: The Causes and Consequences of DBCP-Induced Damage in Costa Rica and Beyond." *International Journal of Health Services* 21, no. 4: 731–57.

———. 1996. "New Harvests, Old Problems: The Challenges Facing Latin America's Agro-export Boom." In *Green Guerrillas: Environmental Conflicts and Initiatives in Latin America and the Caribbean,* edited by H. Collinson, 122–31. London: Latin American Bureau.

Thrupp, Lori Ann, Gilles Bergeron, and William F. Waters. 1995. *Bittersweet Harvests for Global Supermarkets: Challenges in Latin America's Agricultural Export Boom.* Washington, DC: World Resources Institute.

Tisdell, Clement A. 2005. *Economics of Environmental Conservation.* Northampton, MA: Edward Elgar Publishing.

Torres, Hebert. 2002. *Manual de las enfermedades más importantes de la papa en el Perú.* Lima, Perú: Centro Internacional de la Papa.

Tosi, Joseph A., Jr. 1969. *República de Costa Rica: Mapa ecológico según la clasificación de zonas de vida del mundo de L. R. Holdridge.* San José, Costa Rica: Centro Científico Tropical and Instituto Geográfico Nacional.

Trejos, Juan. 1966. *Geografía ilustrada de Costa Rica.* San José, Costa Rica: Trejos Hermanos.

Trivelato, Maria D., and Catharina Wesseling. 1992. "Utilización de plaguicidas en cultivos no tradicionales en Costa Rica y otros países centroamericanos: Aspectos ambientales y de salud ocupacional." In *Exportaciones agrícolas no tradicionales ¿promesa o espejismo? Su análisis y evaluación en el Istmo Centroamericano,* edited by A. B. Mendizábal P. and J. Weller, 163–79. Panama: CADESCA.

Tufte, Edward R. 1983. *The Visual Display of Quantitative Information.* Cheshire, CT: Graphics Press.

———. 1995. *Envisioning Information.* 5th ed. Cheshire, CT: Graphics Press.

Tulachan, Pradepp M. 2001. "Mountain Agriculture in the Hindu Kush–Himalaya: A Regional Comparative Analysis." *Mountain Research and Development* 21, no. 3: 260–67.

United States Department of Agriculture Foreign Agricultural Service. 2005. "U.S. Trade Imports: FATUS Commodity Aggregations," United States Department of Agriculture Foreign Agricultural Service, http://www.fas.usda.gov/ustrade/USTIm-Fatus.asp?QI=.

Valdez Salas, Benjamin, Eva I. García Durán, Juan M. Cobo Rivera, and Gustavo López Badilla. 2000. "Impacto de los plaguicidas en la salud de los habitantes del Valle de Mexicali, México." *Revista de Ecología Latinoamericana* 6, no. 3: 15–21.

Valverde G., Edgar. 1986. "Análisis de objetivos y recomendaciones generales." In *Incremento de la exportación y alimentación costarricense a través del mejoramiento del cultivo del chayote*, edited by E. Valverde G., 195–98. San José, Costa Rica: Consejo Nacional de Investigaciones Científicas y Tecnológicas (CONICIT).

Valverde G., Edgar, Elizabeth Carazo Rojas, and Lizbeth Araya Rojas. 2001. *Manipulación, consumo y residuos de plaguicidas en hortalizas y frutas*. San José, Costa Rica: Organización Panamericana de la Salud (OPS).

van den Bosch, Robert. 1980. *The Pesticide Conspiracy*. Garden City, NY: Anchor.

van der Hoek, W., F. Konradsen, K. Athukorala, and T. Wanigadewa. 1998. "Pesticide Poisoning: A Major Health Problem in Sri Lanka." *Social Science & Medicine* 46, no. 4–5: 495–504.

Van Orman, Jan R. 1995. "Mapping the Roots of Sustainable Development in Costa Rica." *Grassroots Development* 19, no. 1: 42–45.

Vargas, Edgar. 1988. "La vejiga del fruto, una nueva enfermedad del chayote (*Sechium edule* L.)." *Agronomía Costarricense* 12, no. 1: 123–26.

———. 1991. "Chayote: *Sechium edule* (Jacq.) Swartz, Cucurbitaceae." In *Aspectos técnicos sobre cuarenta y cinco cultivos agrícolas de Costa Rica,* 327–36. San José, Costa Rica: Ministerio de Agricultura y Ganadería, Dirección General de Investigación y Extensión Agrícola.

Vianna, J. L., and K. Wierer. 1971. *Cartago Beef Packing Company*. Rome: Food and Agriculture Organization.

Vine, Marilyn F., Leonard Stein, Kristen Weigle, Jane Schroeder, Darrah Degnan, Chui-Kit J. Tse, and Lorraine Backer. 2001. "Plasma 1,1-dichloro-2,2-bis(*p*-chlorophenyl)ethylene (DDE) Levels and Immune Response." *American Journal of Epidemiology* 153, no. 1: 53–63.

von Braun, Joachim, David Hotchkiss, and Maarten Immink. 1989. *Nontraditional Export Crops in Guatemala: Effects on Production, Consumption, and Nutrition*. Washington, DC: International Food Policy Research Institute.

von Düszeln, Jürgen. 1991. "Pesticide Contamination and Pesticide Control in Developing Countries: Costa Rica, Central America." In *Chemistry, Agriculture, and the Environment*, edited by M. L. Richardson, 410–28. Cambridge, UK: Royal Society of Chemistry.

Walker, Peter A. 2005. "Political Ecology: Where Is the Ecology? *Progress in Human Geography* 29, no. 1: 73–82."

Ward, Neil. 1995. "Technological Change and Regulation of Pollution from Agricultural Pesticides." *Geoforum* 26, no. 1: 19–33.

Wargo, John. 1998. *Our Children's Toxic Legacy: How Science and Law Fail to Protect Us from Pesticides.* 2nd ed. New Haven, CT: Yale University Press.

Wasilwa, L. A., J. C. Correll, and T. E. Morelock. 1995. "Phytophthora Blight of Squash Caused by *Phytophthora capsici* in Arkansas." *Plant Disease* 79, no. 11: 1188.

Watts, Michael. 1983a. "Hazards and Crisis: A Political Economy of Drought and Famine in Northern Nigeria." *Antipode* 15, no. 1: 24–34.

———. 1983b. *Silent Violence: Food, Famine, and Peasantry in Northern Nigeria.* Berkeley: University of California Press.

———. 1992. "Living under Contract: Work, Production, Politics, and the Manufacture of Discontent in a Peasant Society." In *Reworking Modernity: Capitalisms and Symbolic Discontent,* edited by A. Pred and M. Watts, 65–105. New Brunswick, NJ: Rutgers University Press.

Watts, Michael, and David Goodman. 1997. "Agrarian Questions: Global Appetite, Local Metabolism: Nature, Culture, and Industry in Fin-de-siècle Agro-Food Systems." In *Globalising Food: Agrarian Questions and Global Restructuring,* edited by D. Goodman and M. Watts, 1–32. New York: Routledge.

Watts, Michael, and Richard Peet. 2004. "Liberating Political Ecology." In *Liberation Ecologies: Environment, Development, Social Movements,* edited by R. Peet and M. Watts, 3–47. New York: Routledge.

Weir, David. 1987. *The Bhopal Syndrome: Pesticides, Environment, and Health.* San Francisco: Sierra Club Books.

Weir, David, and Mark Schapiro. 1981. *Circle of Poison: Pesticides and People in a Hungry World.* San Francisco: Institute for Food and Development Policy.

Wells, Miriam J. 1996. *Strawberry Fields: Politics, Class, and Work in California Agriculture.* Ithaca, NY: Cornell University Press.

Wesseling, Catharina, Anders Ahlbom, Daniel Antich, Ana Cecelia Rodríguez, and Roberto Castro. 1996. "Cancer in Banana Plantation Workers in Costa Rica." *International Journal of Epidemiology* 25, no. 6: 1125–31.

Wesseling, Catharina, Daniel Antich, Christer Hogstedt, Ana Cecelia Rodríguez, and Anders Ahlbom. 1999. "Geographical Differences of Cancer Incidence in Costa Rica in Relation to Environmental and Occupational Pesticide Exposure." *International Journal of Epidemiology* 28, no. 3: 365–74.

Wesseling, Catharina, Marianela Corriols, and Viria Bravo. 2005. "Acute Pesticide Poisoning and Pesticide Registration in Central America." *Toxicology and Applied Pharmacology* 207 (2, Supplement 1): 697–705.

Whatmore, Sarah. 2002. *Hybrid Geographies: Natures, Cultures, Spaces.* Thousand Oaks, CA: Sage.

Wheat, Andrew. 1996. "Toxic Bananas." *Multinational Monitor* 17, no. 9: 9–15.

Whorton, James C. 1974. *Before Silent Spring: Pesticides and Public Health in Pre-DDT America.* Princeton, NJ: Princeton University Press.

Williams, Robert G. 1986. *Export Agriculture and the Crisis in Central America.* Chapel Hill: University of North Carolina Press.

Williamson, Stephanie. 2003. *Pesticide Provision in Liberalised Africa: Out of Control?* Agricultural Research and Extension Network Paper No. 126. London: Overseas Development Institute.

Wilson, Clevo. 2000. "Environmental and Human Costs of Commercial Agricultural Production in South Asia." *International Journal of Social Economics* 27, nos. 7–10: 816–46.

Wilson, Clevo, and Clem Tisdell. 2001. "Why Farmers Continue to Use Pesticides Despite Environmental, Health and Sustainability Costs." *Ecological Economics* 39, no. 3: 449–62.

Wing, Steve. 2000. "Limits of Epidemiology." In *Illness and the Environment: A Reader in Contested Medicine,* edited by S. Kroll-Smith, P. Brown, and V. J. Gunter, 29–45. New York: New York University Press.

Winter, Michael. 1997. "New Policies and New Skills: Agricultural Change and Technology Transfer." *Sociologia Ruralis* 37, no. 3: 363–81.

Wolf, Steven A., Brent Hueth, and Ethan Ligon. 2001. "Policing Mechanisms in Agricultural Contracts." *Rural Sociology* 66, no. 3: 359–81.

World Health Organization. 1990. *Public Health Impacts of Pesticides Used in Agriculture.* Geneva: World Health Organization.

Worster, Donald. 1990. "Transformations of the Earth: Toward an Agroecological Perspective in History." *Journal of American History* 76, no. 4: 1087–106.

Wright, Angus. 1986. "Rethinking the Circle of Poison: The Politics of Pesticide Poisoning among Mexican Farm Workers." *Latin American Perspectives* 13, no. 4: 26–59.

———. 1990. *The Death of Ramón González: The Modern Agricultural Dilemma.* Austin: University of Texas Press.

Wu, Ming-Ling, Jou-Fang Deng, Wei-Jen Tsai, Jiin Ger, Sue-Sun Wong, and Hong-Ping Li. 2001. "Food Poisoning Due to Methamidophos-Contaminated Vegetables." *Clinical Toxicology* 39, no. 4: 333–36.

Yen, Ivan Chang, Isaac Bekele, and Carlyle Kalloo. 1999. "Use Patterns and Residual Levels of Organophosphate Pesticides on Vegetables in Trinidad, West Indies." *Journal of the Association of Official Analytical Chemists* 82, no. 4: 991–95.

Zaidi, Iqtidar H. 1984. "Farmers' Perception and Management of Pest Hazard." *Insect Science and Its Applications* 5, no. 3: 187–201.

Zimmerer, Karl S. 1991. "Wetland Production and Smallholder Persistence: Agricultural Change in a Highland Peruvian Region." *Annals of the Association of American Geographers* 81, no. 3: 443–63.

————. 1996. *Changing Fortunes: Biodiversity and Peasant Livelihood in the Peruvian Andes.* Berkeley: University of California Press.

————. 1999. "Overlapping Patchworks of Mountain Agriculture in Peru and Bolivia: Toward a Regional-Global Landscape Model." *Human Ecology* 27, no. 1: 135–65.

Zimmerer, Karl S., and Thomas J. Bassett. 2003. "Future Directions in Political Ecology: Nature-Society Fusions and Scales of Interactions." In *Political Ecology: An Integrative Approach to Geography and Environment-Development Studies,* edited by K. S. Zimmerer and T. J. Bassett, 274–95. New York: Guilford.

Zúñiga, Claudia, Irma Maroto, Fernando Ramírez, and José Rodríguez. 1991. "Valoración de colinesterasas en trabajadores agrícolas expuestos a insecticidas organofosforados en el cultivo de la papa." In *El deterioro ambiental en Costa Rica: Balance y perspectivas,* edited by J. Gracia Bondía, 179–82. San José: Editorial de la Universidad de Costa Rica.

Zúñiga, Luis Emilio. 1986. "Aspectos económicos del cultivo del chayote (*Sechium edule* SW)." Unpublished thesis, Economía Agrícola, Facultad de Agronomía, Universidad de Costa Rica, San José, Costa Rica.

Zúñiga, Luis Emilio, Walter González M., and Jorge Fonseca Z. 1986. "Evaluación económica del chayote en Costa Rica." In *Incremento de la exportación y alimentación costarricense a través del mejoramiento del cultivo del chayote,* edited by E. Valverde G., 172–94. San José, Costa Rica: Consejo Nacional de Investigaciones Científicas y Tecnológicas (CONICIT).

Zwankhuizen, Maarten J., Francine Govers, and Jan C. Zadoks. 1998. "Development of Potato Late Blight Epidemics: Disease Foci, Disease Gradients, and Infection Sources." *Phytopathology* 88: 754–63.

Index

acaricides. *See under* pesticides, modes of action of

acetylcholinesterase inhibition, 247–9. *See also* carbamates; pesticides, chemical classes of

agricultural activities: clearing, 85, 90; harvesting, 13, 19, 39, 43, 48, 54, 56–57, 82, 88, 99, 100–101, 108, 112, 114, 134, 139, 141–44, 154, 160, 163–64, 179, 188–90, 199–203, 222–23, 242n3, 249; planting, 55–57, 59, 67, 83, 90, 102, 133, 154, 162, 219; spraying. *See also under* pesticide activities

agricultural census: of Costa Rica, 95–99, 102, 221, 239n2, 242n2

agrochemical input manufacturers and retailers, 24, 59–60, 62, 86, 118, 211, 219, 227, 239n4, 239n5, 240n5

agroecology. *See under* alternative agriculture

agroexports: new agroexports, 10, 13–16, 19, 32–34, 37–38, 48, 113, 117, 119, 125, 147–53, 156, 172–74, 207–10, 226, 243; traditional agroexports, 10, 13. *See also* bananas; coffee; sugarcane

aldicarb. *See under* pesticides, active ingredients of

alternative agriculture: agroecology, 4, 9, 38, 84, 89, 105, 118, 125, 210, 213, 217; inputs of, 3, 133, 186, 192, 194–95, 199, 204–5; organic agriculture,

58, 65, 80, 114–15, 119, 125, 136, 164, 205, 212, 217, 220; mentioned, 10, 65, 119, 174, 206, 216

avermectin. *See under* pesticides, active ingredients of

Bacillus thuringiensis. *See under* pesticides, active ingredients of

bacteriacides. *See under* pesticides, modes of action of

bad actor pesticides, 93, 96–97, 235, 247

bananas, 5, 10, 12, 14, 16, 83, 90–91, 98, 123–24, 213, 242

beans (dried), 83, 90, 95, 98, 106, 224

BHC (lindane). *See under* pesticides, active ingredients of

biopesticides. *See under* pesticides, chemical origin of

Blue Seal, 182–83, 191–92

broccoli, 9, 10, 12, 14, 17, 33, 45–7, 189, 205, 224

cabbage, 9, 12, 14, 17, 45–47, 79, 88, 96, 100, 104–5, 174, 177–80, 189–90, 205, 217, 224, 240

California, 5, 8–9, 23, 175–76, 193, 238n11, 243n4, 247

cancer. *See under* pesticides, health impacts of

capitalism: agrarian capitalism, 4, 10, 19, 37, 41, 58–72, 84–87, 110–19, 207–10, 218; capital accumulation, 35, 43, 60, 65, 68–70, 85–86, 145–48, 157, 207, 239n4;

disease pests (continued)
105; early blight (*Alternaria solani*),
239n3; influence of fertilizer upon,
58–59; late blight (*Phytophthora
infestans*), 52–56, 58, 81–82, 88,
101–5, 107, 111, 205, 228, 239n12,
240n6; mosaic virus, 53, 105; peca
blanca (*Ascochyta phaseolorum*),
113; phytosanitary controls over,
15, 32, 124, 155, 242n5; of potato,
52–56, 58, 81–82, 88, 101–5, 107,
111, 205, 228, 239n3, 239n12,
240n6, 241n4; powdery mildew, 105;
Rhizoctonia solani, 239n3; of squash,
53, 56–57; squash phytophthora
blight (*Phytophthora capsici*), 53,
56–57; vejiga (*Mycovellosiella
cucurbiticola* and *M. lantana*), 112.
See also pest pressure; pesticide
resistance
dithiocarbamates. *See under* pesticides,
chemical classes of
Dominican Republic, 6, 55, 113, 117,
123, 127, 146
endocrine system disruption. *See under*
pesticides, health impacts of
endosulfan. *See under* pesticides, active
ingredients of
environmental geography, 34, 37, 49–58,
68, 72–84, 227–28, 240n2
environmental history, 38, 87, 88–119
Environmental Protection Agency
(United States), 122–24, 129, 139,
146–47, 158, 161–64, 167, 176, 179,
193, 235, 243n, 247
epidemiology, 27, 178, 240n2. *See also*
pesticides, health effects of
EUREPGAP. *See* good agricultural
practices (GAP)
Europe, 3–6, 9, 14, 47, 49, 61, 127, 129,
133, 148, 153, 174–75, 188, 206
export firms, 3, 5, 38, 40, 48–49, 82,
112–15, 119–52, 154–57, 159–60,

161–64, 166, 168, 171, 173, 183–84,
187, 201, 204, 207–9, 220–21, 227,
242n, 243n, 244n
exporters. *See* export firms
exports. *See* agroexports
exposure, to pesticides. *See under*
pesticides, health impacts of
externalities, negative, of agrochemical
use, 22, 64, 154, 216
farmers: export-oriented, 3–4, 33,
38–39, 68, 72–73, 110–15, 146–51,
156–70, 173, 191–97, 200–202, 204,
206, 208, 213, 219–21, 227, 229–31;
farm households and livelihoods,
30, 39, 41–46, 66, 68–72, 77–80,
83–87, 90, 100, 154, 190, 203, 226,
239n4; large-scale, 8, 10, 42–43, 49,
62, 67–68, 83–85, 92, 137, 148, 221,
241n6, 242n8, 243n3; smallholders,
3, 7–8, 10, 18, 41, 43–44, 64, 69,
78, 83–84, 95, 97, 118, 148–49,
154, 183, 191, 202–3, 221, 244n7;
national market-oriented, 90–105,
168, 174–76, 188, 190–204, 206,
213, 220–21
farmworkers, 25–26, 28, 45, 63, 66,
211, 213–15, 248
fertilizers, 47, 64; chemical/synthetic, 10,
21, 23, 38, 43, 58–59, 62, 64, 83, 88,
91, 99–101, 103, 105, 107, 112, 133,
206, 239n5; organic, 100, 133
foliar nutrients, 21, 26, 101, 205, 239n5
Food and Agriculture Organization
(United Nations), 5–7, 109, 122, 128,
134, 176, 219
Food and Drug Administration (United
States), 113–14, 122–23, 125–27,
129, 139, 147, 156, 158, 162, 164,
167, 209, 219, 243n2, 243n5
food consumption: of domestic goods
in Costa Rica and developing
countries, 9–10, 16–18, 242n1;
farmers' avoidance of cabbage, 190;

markets requirements (continued)
  export, 35–36, 59, 116, 121, 123–24,
  131, 135, 147, 152, 155, 208, 220;
  mentioned, 30
Material Safety Data Sheet, 193, 247
maximum residue limits (MRLs). *See*
  tolerances
methamidophos. *See under* pesticides,
  active ingredients of
minisquash, 12, 48, 56–57, 67–68,
  72, 74, 76–82, 114, 130, 132, 152,
  159–63, 167, 225–26, 230–31, 241
Ministerio de Agricultura y Ganadería
  (Costa Rica), 7, 17, 110, 130,
  141–42, 144–45, 156, 166, 182–84,
  191, 199; and Sandid Vegetal unit,
  182–84, 191, 199
Ministerio de Salud of Costa Rica, 130
minivegetables, 48–49, 67–68, 79, 110,
  114–15, 121, 126, 128, 130–37, 140,
  145–47, 149, 159–60, 164, 187, 206,
  220, 242, 244
nematicides. *See under* pesticides, modes
  of action of
neurotoxicity. *See under* pesticides,
  health impacts of
onion, 9, 12, 14, 17, 41, 45–7, 79, 96,
  104, 117, 174, 190, 217, 224
organic agriculture. *See under* alternative
  agriculture
organochlorines. *See under* pesticides,
  chemical classes of
organophosphates. *See under* pesticides,
  chemical classes of
paraquat. *See under* pesticides, active
  ingredients of
parts per million (as a measure of
  pesticide residues and dose), 122,
  176, 179, 180, 188–89, 248
pathogens. *See* disease pests
permethrin. *See under* pesticides, active
  ingredients of

pest pressure, 19, 62–63, 83, 114, 118,
  167, 203; geographic variability of,
  49–58, 67–68, 76–77, 227–28. *See
  also* disease pests; insect pests; spatial
  strategy
Pesticide Action Network, 193, 247
pesticide intensity: and socioeconomic
  differentiation, 67–87; calculation
  and measurement of, 5, 205, 223–26,
  249; controlled comparisons of,
  11–12, 72–84, 100, 237n4; export
  orientation influences upon, 8–11,
  33; increases in, 103–4; national
  comparisons of, 6–7, 9, 237n4;
  of small scale farmers, 10, 18, 78,
  83–84. *See also* cosmetic standards
  for produce
pesticide manufacturers and
  formulators, 25–27, 56, 59–62, 86,
  125. *See also* capital
pesticide poisoning. *See under* pesticides,
  health impacts of
pesticide regulation: double movement
  (conceptualized by Karl Polanyi), 4,
  116, 208; double standard concerning
  residues, 171, 173–82, 199, 208;
  enforcement by capital; regulatory
  risk, 28, 152–59, 163–64, 167, 173,
  203, 208; tolerances (MRLs), 121–24,
  128–29, 131, 139, 147, 158, 161,
  163–64, 167, 176, 179, 184, 199,
  235, 243n3, 245n5, 249; violations
  of, 38–39, 113–14, 119–21, 123,
  125–32, 136–37, 139, 143–50,
  155–58, 160, 162–64, 166–68, 181,
  183–84, 198, 209, 215, 244n11.
  See also Blue Seal; Environmental
  Protection Agency; export firms; Food
  and Drug Administration; HortiFruti;
  Ministry of Agriculture
pesticide resistance, development of, by
  pests, 29, 64, 86, 240n6, 244n5

About the Author

Ryan Galt is an associate professor in the Community and Regional Development unit of the Department of Human Ecology at the University of California, Davis. He is also a provost fellow of the Agricultural Sustainability Institute at UC Davis. He received his PhD and MSc in geography from the University of Wisconsin–Madison and his AB (summa cum laude) in geography from the University of California, Berkeley. His research focuses on the relationships between society and environment with an emphasis on food systems and agriculture and has appeared in numerous journals, including *Economic Geography, Global Environmental Change, Food Policy, Human Ecology, Geoforum, Agriculture and Human Values, Annals of the Association of American Geographers,* and *Transactions of the Institute of British Geographers.* His current research utilizes the perspectives of political ecology and the political economy of agriculture to understand a number of topics, including governance of pesticides in industrial agriculture and the development of pest control alternatives, movements aimed at fostering a more just and environmentally sustainable food system, the socioecological sustainability of agricultural and food systems in the context of broader political and economic systems, and the expansion and performance of alternative forms of producer-consumer relationships, including the tensions around realizing social, economic, and ecological values in community supported agriculture. Galt teaches classes in geography, community and regional development, and sustainable agriculture and food systems. His work in Costa Rica was supported by a Fulbright IIE Student Fellowship and a Global Studies Fellowship from the MacArthur Foundation and the University of Wisconsin–Madison International Institute.